PRINCIPLES OF HYDRATION

FLINTS LAWS:

A FOUNDATION FOR THE UNIFICATION

OF THE PHYSICAL SCIENCES

STUART HALE SHAKMAN

InstituteOfScience.com

For information please
contact:

S. H. Shakman
INSTITUTE OF SCIENCE
InstituteOfScience.com
mail@instituteofscience.com
First Printed 2014

PREFACE

This work seeks to supplement and further validate the great works of Lewis Herrick Flint on Hydration as reflected in three articles in the *Washington Academy of Sciences*, 1932-3, and two books,

Rights to reprint the Flint works referred to above have been obtained from the respective publishers. See *Behavior Patterns of Hydration* 1964, and *Hydration and Biology*, 1968, available through Amazon.com.

* * * * * * *

"Every sound distinction, illuminating definition, and experimental result, with which we have become acquainted becomes, as soon as it has been accepted, 'obviously true'.

"It will be remembered in this connexion that ordinary proverbial philosophy insists that every novel ideal or new invention must, before it wins general acceptance, pass through three stages.

-- It is, to begin with, repudiated as 'absurd'.

-- After that, it is allowed to be 'reasonable'.

-- And finally it is belittled as 'obvious'."

Almroth E. Wright, MD, ScD, FRS, *Prolegomena to the Logic Which Ssearches for Truth* (1941).

****

"Some people without brains do an awful lot of talking."

The Tin Man, Wizard of Oz.

****

"For the time being, we have to admit that we do not possess any general theoretical basis for physics which can be regarded as its logical foundation. The field theory, so far, has failed in the molecular sphere. ...

"Some physicists, among them myself, cannot believe that we must abandon, actually and forever, the idea of direct representation of physical reality in space and time; or that we must accept the view that events in Nature are analogous to a game of chance. ..."

Albert Einstein
Nature **345**, 6 276 (14 June 90), from Nature June 15, 1940

CONTENTS

HYDRATION ALGEBRAICS

Herein lies the secret code of the universe -- the essential key to the unification of the physical sciences -- and the ultimate computer game. (This is not a computer game in the conventional sense. It is certainly computer friendly, and there are prizes ahead. But the game is very real, as it deals with the very fabric of the universe; and the prizes are not the usual trival type - think Pulitzer and Nobel.)

And we owe it all to the brilliant discoveries and years of dedication of Lewis Herrick Flint. Flint toiled on his studies of hydration for four decades in virtual obscurity (he was a well-respected botanist), and his work documents beyond all doubt his (algebraic) description of hydrational potentiality.

Flint's great works will come to place him among the greatest and most important scientists of all time, on a par with the likes of Johannes Kepler, Aristarchus of Samos, and precious few others.

Remember, as the National Science Foundation says:

"SCIENCE IS FUN".

S. H. Shakman, INSTITUTE OF SCIENCE

Shakman, S. H., Nature 338, 456 (1989):

Heliocentric tangents

SIR -- The heliocentric hypothesis, so able championed by Copernicus and Galileo(1), is authoritatively said to have originated with Aristarchus of Samos in the third century BC. Sand-Reckoner, written by Aristarchus's younger contemporary Archimedes before 216 BC(2), attributed to Aristarchus a book containing the hypotheses "that the fixed stars and the sun remain unmoved, [and] that the earth revolves about the sun in the circumference of a circle, the sun lying in the middle of the orbit...". Aristarchus's (lost) book is thought to have "clearly...also included some kind of geometrical proof"(3).

Aristarchus had also produced a treatise On the sizes and distances of the Sun and Moon, which has survived intact. Its "excellent" methodology confirms that Aristarchus's (probably later) heliocentric hypothesis was similarly "not irresponsible" but rather the work of a conscientious astronomer"(2).

Nicholas Copernicus's De Revolutionibus (1543) acknowledged Aristarchus but not his heliocentric hypothesis (4); however, it seems certain that Copernicus was also acquainted with the latter. For example, his original manuscript had referred to the opinion of Aristarchus on the movement of the Earth, but this reference was subsequently "suppressed" or "scored out" (4). Moreover, in relating the views of Philolaus, Heracleides and Ecphantus on the question of movement of the Earth, Copernicus's Preface quoted from De Placitis Philosophorum of pseudo-Plutarch, a work in which may also be found: " Aristarchus places the Sun among the fixed stars, and holds that the Earth moves around the Sun's circle"(4). And De Revolutionibus (IV, 32) cites Archimedes' Measurement of the Circle, a treatise commonly found in the company of Sand-Reckoner(5).

Copernicus's unquestionably pivotal contribution to astronomy was his grand revival of the heliocentric hypothesis as a systematic planetary theory(4). But in order to fit theory to observation, Copernicus had retained the geometric devices used by Ptolemy (the deferent, epicycle and excentric), and had referred details of planetary movements not to the Sun but rather to the centre of the Earth's orbit. Because of technical and other difficulties with the copernican system, the astronomer Tycho Brahe (1546-1601) rejected it. Brahe had compiled an unrivalled set of observations, which he thought would demonstrate that the Sun and Moon travel around the Earth while the other planets travel round the Sun(6).

After Brahe's death, Johannes Kepler (1571-1630) invested years in the analysis of Brahe's data, culminating in the derivation of Kepler's three laws of planetary motion. These provided a precise and enduring mathematical characterization of the heliocentric hypothesis, thus serving to support the position of Copernicus while ironically refuting that of Brahe.

STUART HALE SHAKMAN

1. Nature 337, 101 (1989).
2. Sarton, G. A History of Science: Hellenistic Science and Culture in the Last Three Centuries BC, 54-57 (Harvard University, Cambridge, 1959).
3. Heath, T. Aristarchus of Samos, 301-302 (Clarendon, Oxford, 1913).
4. Armitage, A. Copernicus (Allen & Unwin, London, 1938).

5. Heath, T., The Works of Archimedes xxiv-xxvi (Dover, N.Y., 1912).
6. Armitage, A. Sun Thou Stand Still 149, 169-175 (Sigma, London, 1947).

CORRESPONDENCE

Heliocentric tangents

SIR—The heliocentric hypothesis, so ably championed by Copernicus and Galileo[1], is authoritatively said to have originated with Aristarchus of Samos in the third century BC. *Sand-Reckoner*, written by Aristarchus's younger contemporary Archimedes before 216 BC[2], attributed to Aristarchus a book containing the hypotheses "that the fixed stars and the sun remain unmoved, [and] that the earth revolves about the sun in the circumference of a circle, the sun lying in the middle of the orbit . . .". Aristarchus's (lost) book is thought to have "clearly . . . also included some kind of geometrical proof"[3].

Aristarchus had also produced a treatise *On the sizes and distances of the Sun and Moon*, which has survived intact. Its "excellent" methodology confirms that Aristarchus's (probably later) heliocentric hypothesis was similarly "not irresponsible" but rather the work of a "conscientious astronomer"[3].

Nicholas Copernicus's *De Revolutionibus* (1543) acknowledged Aristarchus but not his heliocentric hypothesis[4]; however, it seems certain that Copernicus was also acquainted with the latter. For example, his original manuscript had referred to the opinion of Aristarchus on the movement of the Earth, but this reference was subsequently "suppressed"[4] or "scored out"[4]. Moreover, in relating the views of Philolaus, Heracleides and Ecphantus on the question of movement of the Earth, Copernicus's Preface quoted from *De Placitis Philosophorum* of pseudo-Plutarch, a work in which may also be found: "Aristarchus places the Sun among the fixed stars, and holds that the Earth moves around the Sun's circle"[4]. And *De Revolutionibus* (IV, 32) cites Archimedes' *Measurement of the Circle*, a treatise commonly found in the company of *Sand-Reckoner*[5].

Copernicus's unquestionably pivotal contribution to astronomy was his grand revival of the heliocentric hypothesis as a systematic planetary theory[4]. But in order to fit theory to observations, Copernicus had retained the geometric devices used by Ptolemy (the deferent, epicycle and excentric), and had referred details of planetary movements not to the Sun but rather to the centre of the Earth's orbit. Because of technical and other difficulties with the copernican system, the astronomer Tycho Brahe (1546–1601) rejected it. Brahe had compiled an unrivalled set of observations, which he thought would demonstrate that the Sun and Moon travel around the Earth while the other planets travel round the Sun[6].

After Brahe's death, Johannes Kepler (1571–1630) invested years in the analysis of Brahe's data, culminating in the derivation of Kepler's three laws of planetary motion. These provided a precise and enduring mathematical characterization of the heliocentric hypothesis, thus serving to support the position of Copernicus while ironically refuting that of Brahe.

STUART HALE SHAKMAN
2269 Chestnut St, Apt 115,
San Francisco, California 94123, USA.

1. *Nature* **337**, 101 (1989).
2. Sarton, G. *A History of Science: Hellenistic Science and Culture in the Last Three Centuries* BC, 54–57 (Harvard University, Cambridge, 1959).
3. Heath, T. *Aristarchus of Samos*, 301–302 (Clarendon, Oxford, 1913).
4. Armitage, A. *Copernicus* (Allen & Unwin, London, 1938).
5. Heath, T. *The Works of Archimedes* xxiv-xxvi (Dover, New York, 1912).
6. Armitage, A. *Sun Thou Stand Still* **140**, 169–175 (Sigma, London, 1947).

Subject: GROUNDS FOR COPERNICAN CONNIPTION

SIR -- Although later work concerning stellar aberration and parallax may have further validated the heliocentric hypothesis[1], indisputable and generally accepted proof (i.e., enduring "laws") had in fact been established during Galileo Galilei's lifetime by his pen-pal Johannes Kepler. In 1609 Kepler correctly described planetary orbits as elliptical, and ten years later he delivered this stunning declaration of their mathematical precision[2]:

> " ... at last, at last the right ratio of the periodic times to the spheres ... outfought the darkness of my mind by the great proof afforded by my labor of seventeen years on Brahe's observations and meditation upon it uniting in one concord, in such fashion that I first believed I was dreaming and was presupposing the object of my search among the principles. But it is absolutely certain and exact that <u>the ratio which exists between the periodic times of any two planets is precisely the ratio of the 3/2-th power of the mean distances</u> ... ."

And while Kepler asserted that Copernicus "did not have the slightest inkling" of the prior heliocentricity of Aristarchus, unaware that Copernicus's acknowledgment of debt to Aristarchus had been scored out of the original <u>De Revolutionibus</u> manuscript prior to printing[3], similarities between Copernicus's earlier <u>Commentariolus</u>[4] and Archimedes's <u>Sand-Reckoner</u> characterization of Aristarchus's views[5] are nonetheless noteworthy:

Aristarchus, per Archimedes: "His hypotheses are that <u>the fixed stars and the sun remain unmoved, that the earth revolves about the sun in the circumference of a circle, the sun lying in the middle of the orbit,</u> and that the sphere of the fixed stars, situated about the same centre as the sun, is so great that the circle in which he supposes the earth to revolve bears such a proportion to the distance of the fixed stars as the center of the sphere bears to its surface."

Copernicus, in <u>Commentariolus</u>: "All the spheres revolve about the sun as their mid-point and therefore the sun is the center of the universe. ... The ratio of the earth's distance from the sun to the height of the firmament is so much smaller than the ratio of the earth's radius to its distance from the sun that the distance from the earth to the sun is imperceptible in comparison with the height of the firmament."

Here, Copernicus's failure to cite Aristarchus might be explained by the fact that this preliminary sketch (the so-named <u>Commentariolus</u>) was not published, and may not even have been titled, by Copernicus[4]. And as for deletion of reference to Aristarchus's heliocentricity from <u>De Revolutionibus</u>, attributed by some to Copernicus[3], more logically this might have been the work of its unintended final editor, Andreas Osiander. Such a deletion would have worked not to the purpose of Copernicus, who viewed heliocentricity as reality, but rather to that of Osiander, whose unauthorized preface to <u>De Revolutionibus</u>, inserted anonymously, characterized the Copernican scheme as a device for calculation.

It is known that Copernicus received the printed <u>De Revolutionibus</u> on the day of his death. Perhaps the sight of unauthorized alterations contributed to the precise timing of his departure.

Stuart Hale Shakman, P. O. Box 382, Santa Monica, CA., U.S.A. 90406-0382
              16 June 1993; submitted to Nature; not printed.

1. Psimopoulos, M. and T. Theocharis, Nature **363** (1993), 108.

 2. Kepler, Johannes, Harmonies of the World, 1619, Chapter 3, VIII   transl. by C. B. Wallis (U. of Chicago/Encyclopaedia   Britannica, 1952).

 3. Copernicus, Nicholas, On the Revolutions (De Revolutionibus),   trans. by E. Rosen (Johns Hopkins University Press, Baltimore,   1978), pp. 25, 334-5, 360.

 4. Archimedes, The Sand-Reckoner, in Heath, T. L., The Works of   Archimedes (Dover, New York, 1912), pp. 221-2.

 5. Copernicus, Nicholas, Nicolai Copernici de hypothesibus motuum   caelestium a se constitutis commentariolus or Commentariolus,   transl. by E. Rosen in Three Copernican Treatises (Columbia   University Press, 1939), pp. 6, 22-4, 58.

LOGIC, TECHNOLOGY AND THE UNIFICATION OF SCIENCE

--- The Logic of Hydration as Key to the Unification of Science
--- Required Technology Advances: Wheatstone Bridge; Slide Rule

Neither Flint nor this writer was specifically looking for an understanding of or key to the unification of the physical sciences. Nonetheless, this unsought entity seems indeed to have been located, as herein discussed. Moreover in retrospect, if one were looking for such a unifying principle, hydration would seem to be a preeminently logical place to begin.

The "aqueous" (water) environment is home to vital investigations by biologists, chemists and physicists alike. Without water: there is no life and thus no biology; those test-tube chemistry experiments could not begin; etc.

The term "hydrate" refers to a combination of water with some other substance; hydrates are known to exist in aqueous (water) solutions, in crystals, and in more complex forms, e.g., carbohydrates. And when a carbohydrate (e.g. an apple or a person) is deficient in water, it is said to be "dehydrated", which situation might be corrected by adding a sufficient quantity of water. But the very existence of a precise mechanism whereby these carbo- or other hydrates, even simple aqueous solutions, get to be hydrated in the first place is not generally recognized. It would not have been recognized at all, at least not within our lifetimes, were it not for two marvelous inventions of mankind - the Wheatstone bridge, and the slide rule.

WHEATSTONE BRIDGE: The Wheatstone bridge (or "Wheatstone's bridge") is a device that measures conductance of electricity through aqueous solutions. It is named for the reknowned Sir Charles Wheatstone, inventor of the automatic telegraph; however, Wheatstone did not invent the Wheatstone bridge. Rather, it was devised by S. H. Christie in 1833; following Wheatstone's use of it in 1847 (Scientific Papers, p. 129 and Phil. Trans. 1847), it came to be associated with his name.

By the late 19th Century, the means of dividing total conductance of a given aqueous (salt) solution ("electrolyte") into components attributable to the separate ions was known (Kohlrausch, Arrhenius, etc.); and it was also known: (1) that the conductance values for these ions also comprise a direct and precise measure of their relative mobilities (but not generally emphasized, nor is it today); (2) that these mobilities are some function of the weight of the hydrated ions (Abegg and Bodlander, 1899); (3) that, in the case of diffusion of gases, mobilities vary with the inverse-square-root of weight (Graham's law) and (4) that the overall behavior of solute-ions and gases were analogous (van't Hoff).

In 1932, Lewis H. Flint was able to put it all together, with the indispensable assistance of that modern marvel, the slide rule.

THE MARVELOUS SLIDE RULE

While the invention of logarithms (by John Napier of Scotland) and early slide rules can be traced back to the early 17th century, a lack of general availability even through the end of the 19th Century seems reflected in Flint's suggestion that Abegg and Bodlander's 1899 work may have been limited due to a lack of a slide rule (or due to fear of rejection). This is a bit unfair, considering the incredibly lucky set of circumstances that presented themselves to Flint and greatly facilitated his discovery of the algebraics of hydration [See UNI-SCIENCE ABSTRACTS "[A] An Algebraic Approach to Unification".]

In any case, Flint did have a slide rule, and this enabled him to readily perform the calculations which explored the implications of characterizing relative conductance/mobility as inverse-square-root of relative (hydrated ionic) weights, and disclosed an inverse, periodic, integral and reciprocal relationship between atomic numbers and hydration numbers. [Flint, L.H., Behavior Patterns of Hydration, 1964; see also "A Partial Bibliography". In the absence of reference to Flint's work, modern chemistry employs various theoretical methods in admittedly not- particularly-successful attempts to determine hydration numbers or even gain a basic understanding of hydration. It is noted that Flint's work differs from these in that his work is not theoretical.

Flint's work involves a direct conversion of unquestionably precise conductance measurements into as-unchallengable relative weight values; the Wheatstone bridge is an electronic scale which precisely measures solute ionic (velocity-thus-) weight. Flint merely described what this scale revealed (he learned how to read the "scale"); hence he referred to his work as a "description" of hydrational potentiality. If that sounds too technical to take in the first bite, just think of it as "the secret code of the universe" and a key component of the "mathematics of metabolism, all of which it is, and consider using it to play "the ultimate computer game".

John Napier of Scotland announced the discovery of logarithms in 1614, whose intended purpose was to remove "hindrance and tedium" from calculations; two decades earler, in 1594, Napier had privately comunicated a summary of his results to Tycho Brahe (p2). In 1617 Henry Briggs transformed logarithms to the form described in modern definitions (p3); following this, Edmund Gunter laid off on a strip of wood (p4).

Prior to 1630 William Oughtred invented the double rule and the circular rule. In 1657 Seth Partridge devised the duplex slide rule - 3 strips with numbers sliding. As described by F. Stone in 1726, the first recorded use of "runner", moving marking line, was by Newton to solve cubic equations (p7).

In 1880 the Mannheim rule was introduced in the U.S. (p17); in 1891, Cox patented a modified "duplex" slide rule.

Source: Thompson, J.E., A Manual of the Slide Rule, D. van Nostrand Co., N.Y. 1930

WHY FLINT'S WORK IS IGNORED

This is kind of a silly question in a way; most things are ignored, all but those we pay attention to. But in this day and age, what with all the geniuses (or genii) running around, we are compelled to guess how they might all be overlooking THE SECRET CODE OF THE UNIVERSE.

That Flint's work has evaded serious consideration by the science community for several decades is not surprising in that

(1) it poses seemingly irreconcilable differences with conventional wisdom; most notably in the form of a relative mass of 2 for hydrogen on a scale of 16 for oxygen; and

(2) Relatively obscure status and publication in a relatively obscure journal (The Journal of the Washington Academy of Sciences). We may note that Aristarchus of Samos, of the generation before Archimedes, was a leading scientist of his day, witness his starring role in Archimedes' Sandreckoner, nonetheless Aristarchus' Heliocentric hypothesis went begging for some 1800 years up to its revival by Copernicus and proof by Kepler as particularly embodied in his third harmonic law, describing the constancy of the relation of the planets to the sun, i.e. the ratios of the cubes of their distances (D) from the sun and the squares of the times (T) of their revolutions around the sun: $D^3 / T^2 = Constant$

METHODOLOGY AWAITING TECHNOLOGY

While Flint's "model" was derived without benefit of a computer or even a hand-calculator, he gratefully acknowledged the invaluable role of the slide rule in simplifying otherwise cumbersome calculations. He even mused that the complexity of the task and possible lack of a suitable calculating instrument might have discouraged Abegg and Bodlander from realizing the logical consequence of their 1899 work (i.e., the methodology reported by Flint). [Flint1964].

FLINT'S COUP

While subsequent investigators including Walter Nernst [Nernst, Walter, *Theoretical Chemistry*, MacMillan and Co., Ltd., London 1916, p. 397] similarly discussed the concept of an inverse relationship between mobility and hydration (as do some contemporary works [e.g., Robinson, FNH, in 1988 Encyclopedia Brittanica, Vol. 18, . 268]), Flint was certainly the first, and apparently still the only, one to characterize it mathematically as "inverse, integral, reciprocal and periodic" (Flint 64).

Please do not confuse this with the so-called current crop of "grand unification theories". This is a simple algebraic system, based on classical mechanical principles dating back to Newton and earlier, that actually seems to work - - that actually seems to describe the behavior of ions in solution.

Hawking [p. 8, A Brief History of Time] notes the simplicity and limits of "Aristotle's theory that everything was made out of four elements, earth, air, fire and water", in contrast to the predictive advantage of "Newton's theory of gravity … in which bodies attract each other with a force that was proportional to … their mass and inversely proportional to the square of the distance between them." Thus the Newtonian theory "predicts the motions of the sun, the moon, and the planets to a high degree of accuracy", but it was Flint who was first able to apply the inverse squares principle, so prominent in Newton and forebears' works, to the Aristotelian element of water in all of its forms and combinations.

THE BIGGER PICTURE?

As self-appointed custodian of the Secret Code of the Universe, it is difficult for one so closely involved as I to objectively assess the implications of Flint's discovery. This is the fabled unification of science, clothed in beautiful symmetrical implosions, the secret code of the universe including the mathematics of metabolism, a handle to turn on the magical fountains in Ponce de Leon's premonitions.

And now we get to look into that same magical future, but with the knowledge that it's really mathematically certain to happen, if not for our children perhaps for theirs. It could happen that fast. And not just from wishing for it but by figuring out how to make it happen ourselves. All the prophets saw it happening in their visions –but algebra hadn't been figured in.

So perhaps in this marriage of mathematics and science we also find the unification of science, mathematics and religion as well, leading to a harmonious world which we are mathematically destined to master, if we are able to survive our petty aggressions. (We are mindful that the logical extension of religion, the transcending of physical limitations through powers of mind, also unites religion and science; see "The Matter Of Time" available through Amazon.com).

At the same time we are reminded of the "bigger picture" perspective of Gustave Lebon, 1907, asserting that "Worlds peopled like ours, covered with flourishing cities filled with the marvels of science and the arts, must have emerged from eternal night and returned thereto without leaving a trace behind them." [Gustave Le Bon, *Evolution of Forces* 1907, p. 83; see Appendix B]. So might our civilization be the one to transcend the eons and leave an enduring legacy for all time and all space? However unlikely, such is the hope and the inspiration that fuels our seemingly insatiable and enduring quest for an understanding of our place of our universe.

Lewis Herrick Flint – A Partial Bibliography

1926, January, "Electroculture", by L.J. Briggs, A.B. Campbell, R.H. Heald, and L.H. Flint, Office of Biophysical Investigations, Bureau of Plant Industry in U.S. Department of Agriculture, Department Buletin No. 1379, pp. 17-32. San Francisco City Library (SFCL)#630.6 Un 3b56

1928 "Crop Plant Stimulation with Paper Mulch", SFCL#630.63 Un 3 t-1929 "Suggestions for Paper Mulch Trials", 8pp. SFCL#630.62 Un 3 ci

1928 L.H.Flint, G.N. Collins, "Electrical Stimulation of Plant Growth," pp. 585-600. SFCL#630.6 Un 3938

1932 March 4, JOURNAL OF THE WASHINGTON ACADEMY OF SCIENCES (JWAS), Vol. 22, No.5 (March 4, 1932) pp. 97-119, "Hydration of the solute ions of the lighter elements".

1932, April 19, JWAS, Vol. 22, No. 8 (April 19, 1932), pp. 211-217, "The hydration of the solute ions of the heavier elements".

1932, May 24, JWAS,Vol. 22, No. 9 (May 24, 1932), pp. 233-237, "Unhydrated solute element ions".

1932 BOBOLINKS, L.C.#P.S. 3511 .l68B6.

1934, January: "Hydration of Solute Ions in Relation to Acidity, Alkalinity, and PH", Plant Physiol. 1934 January; 9(1): 107–126.

1934 LOLLY-POP PEOPLE, L.C.#PS3511.L68L6

1934, July 13, "Light in Relation to Dormancy and Germination in Lettuce Seed", SCIENCE, p. 39.

1934, October 2, "Sensitivity of dormant lettuce seed to light and temperature" (author's abstract), Proceedings, Botanical Society, 259th meeting, in JOURNAL OF THE WASHINGTON ACADEMY OF SCIENCES, Vol. 25, No. 2, pp. 96-97.

1935, June 24, "Wave Lengths of Radiation in the Visible Spectrum Inhibiting the Germination of Light-Sensitive Lettuce Seed", Smithsonian Miscellaneous Collections, Vol. 94, No. 5, Publication 3334.

1937, June 16, "Wave Lengths of Radiation in the Visible Spectrum Promoting the Germination of Light-Sensitive Lettuce Seed", Smithsonian Miscellaneous Collections, Vol. 96, No. 2, Publication 3414.

1939, April, L.H.F. and Charles F. Moreland, "Response of Lettuce Seedlings to 7600A Radiation", AMERICAN JOURNAL OF BOTANY, Vol. 26., pp. 231-233.

1964 BEHAVIOR PATTERNS OF HYDRATION, Institute for the Advancement of Science and Culture, New Delhi.

1968 HYDRATION AND BIOLOGY, Institute for the Advancement of Science and Culture, New Delhi.

1968 FRESHWATER RED ALGAE OF NORTH AMERICA, Vantage Press, L.C.#QK 569.R4F4

1973 DISSENTING APE

1973 BOTANIST BECALMED

LEWIS HERRICK FLINT – BIOGRAPHY

Lewis Herrick Flint
Who's Who in America
Vol. 28, 1954-5

Flint, Lewis Herrick, scientist; b. Milton, Vt., F3b. 5, 1893; s. Hale Leroy and Anna Martin (Herrick) F.; BS U. of Vt., 1915, MS 1916; student, Harvard Grad School 1918-9; PhD U. of Vt. 1923; m. Jessie Marguerite Chedel, April 28, 1917; children Alden C., Marguerite Ann (Mrs. Wm. E. Holden), Christine Isabel (Mrs. A.N. Robinson, Jr.; 2nd husband Fogelman), Austin W. Instructor botany and assistant botanist, U. of Vt. and Vt. Exptl. Station 1916-1923; Austin Teaching Fellow, Harvard 1918-9; assoc plant physiologist, Bureau of Plant Industry, U.S. Dept. of Agriculture, Washington, DC 1923-1936; plant physiologist, Boyce Thompson Institute for Plant Research, Inc., 1936-7; prof Botany LSU since 1937. Served with US Army WWI. Mem. AAAS, Botany Soc of America, American Soc of Plant Physiologists.

Added information in *American Men of Science*, 1965 (fields of study):

Paper mulch; electro-culture; hydration of ions; light-sensitive seed; quality of light and plant growth; freshwater red algae.

SIGNIFICANCE OF INVERSE SQUARES

How do people discover "big" things, for example, that the earth travels around the sun? In this instance, Aristarchus of Samos, in the third century B.C., had the good fortune of having Heraclides (?) come before him. Based on his prior observation that the inner planets travel around the sun, Aristarchus took the next logical step of attributing to the earth this same revolutionary motion. Even earlier, in the late 5th Century BC, Philolaus had referred to a central "fire".

More than 1800 years later, Tycho Brahe's precise observations of the positions of the heavenly bodies enabled J. Kepler to formulate his three laws of planetary motion. Of Kepler's first law, that planetary orbits describe ellipses, may we not assume that in some fashion Kepler plotted the points described by Brahe, connected the dots, and saw the elliptical pattern? Of Kepler's second law, that the radii traverse equal areas in equal times, this finding would also rely heavily on accurate data, of the type that did not exist prior to Brahe.

$$r^3 = t^2$$

pi x (average)r-square = area total; p x r-square/t = k

Of Kepler's third harmonic law, he appears to have tried various explanations, as indicated by his apparent euphoria in finding the correct relationship.

Nonetheless, his finding of the correct relationship, that the cubes of the orbital radii vary as the squares of periodic times, is a distinct high point in discovery of the mathematical nature of the universe in which we live, perhaps the most significant since that of Aristarchus.

And since Kepler, what discovery could compare with that which extends the mathematical precision of Aristarcus's and Kepler's universe into the biological realm? Such is the promise of the finding and associated hypotheses of L.H. Flint.   The work of Flint is based on the evolution of an "inverse squares" principle whose initial recognition long preceded Kepler.

In the 13th century, Jordanus de Nemore (around 1239; alive in 1222, 1246; dead by 1260; lived 1200?-1259?) stated: "The velocity of a falling body is observed to be continuously multiplied". (E.A. Moody and M. Clagett, The Medieval Science of Weights, U. of Wisc. Press, 1960, p. 217).

Early in the next century, Thomas of Bradwardine (1290? - 1326+) treated velocity as an instantaneous quality of motion or rate, as distinct from "velocity conceived as total distance traversed per total time elapsed ... [which] pointed ... to the inevitable conclusion that constant forces produce constant accelerations rather than constant velocities." His successors at Merton College ("Calculatores") continued to the deduction of the "correct kinematic law relating time elapsed to distance traversed in a uniformly accelerated motion." [Tractatus de Proportionibus,ed. and dtranslated by H. Lamar Crosby, Jr., U. of Wisconsin Press, Madison, 1955, p. 12

Nicole Oresme, Tractatus de Configurationibus Qualitatum et Motuum (1350), Ed. Marshall Clagett, U. of Wisc. Press 1968, p. 13, in "Introduction", states that "Oresme, against the prevailing opinion, seems to suggest that the speed of hte fall of bodies is directly proportional to the time of fall rather than the distance of fall ... "

Then came Copernicus (1473-1542?) with his Aristarchan revival, Maestlin (whose teacher?), Tycho and then J. Kepler.

p. 107, Introduction: Thomas Hariot, in a piece dated between 1595 and 1605, "does seem to have used the configuration technique and arrived at the relationship of s varies with t-squared.that the velocity of a falling object is "multiplied continuously". In Kepler's time, Thomas Hariot (sp?), between 1595 and 1605 stated that the distance traveled by a falling object varies with the square of the time of the fall, a relationship sometimes attributed to Galilea or others. That the velocity of a falling object thus varies with the square root of the distance of its fall was known by Galileo by 1627, as evidenced in a letter of Gianbattista Baliani. It is generally thought that Torricelli extended this principle to the rate of efflux of fluids, in 1644 (?), with the velocity of efflux being that which would result from a fall from the height of the fluid, and varying with the square root of the height. Thus this relationship is commonly known as Toricelli's theorm on the efflux of fluids, although Boyle had attributed it to Marin Mersenne in 1626.

Huygens in 1673 characterized centrifigal force as (m)V$_2$/d. Around this time, the world was also graced with Newton, Boyle and Hooke. Then the likes of Aepinus and Franklin, Priestly, Coulomb and Cavendish.

In the early 19th century, Prof. John Robison considered the effects of density on the "Torricelli" relationship, noting that for a gas such as air density varies inversely with height, thus velocity of efflux would vary with the inverse square root of density. (Encyclopedia Britannica 1822).

From this Thomas Graham (1805-1869) directly adknowledged the derivation of the principle known as "Graham's law of diffusion", whereby the velocities of efflux for various gases were noted to vary as the inverse square roots of their respective densities. And as the respective relative densities of gases, under identical conditions of pressure and temperature, are also a measure of their rerspective molecular weights, Graham's law has also come to be kknown as describing that the velocity of effusion, or diffusion, of a gas varies with the inverse square root of its molecular weight.

Later, in 1885(?), J. van't Hoff demonstrated how gaseous and solute behavior was analogous, but the data was still lacking to fully utilize this principle in order to develop a viable theory of solute behavior.

The required data set was already in process of developing, this being the measurement of conductivities of various solutes, and Kohlrausch's law of independent migration of ions, which enabled conductivities of the individual ions to be presented.

[In this period, in the late 19th to early 20th centuries, we find the likes of Ostwald, Nernst, Kolhrausch and Arrhenius, then Bousfield and Harry Jones, LeBon and Einstein.]

Even in modern times it is recognized that these conductivity values are among the most reliable measurements available to science, although the precise and fortuitous combination of information and interest and ability in interpreting this information, apparently eluded all investigators of the problem, save one, Lewis Herrick Flint.

In his works, Flint discusses how his hypothesis on hydration was initially derived, and from the information he has given us we may speculate on the precise nature of his discovery. It is a bit mind-boggling to think that one of the most important scientists of all time lived and worked in the 20th century America, died in 1973, and is still unknown.

AN OUTLINE: HISTORY OF THE INVERSE-SQUARE PRINCINCIPLE

A comprehensive discussion would include references to all of the below names encountered in an historical review/ search:

Philolaus - Late 5th Century B.C.; "Central Fire"

Plato Aristotle Ptolemy Pythagoreus

Aristarchus - 3rd century B.C. (Greek); sun at center

Jordanus de Nemore, alive adult 1222-1246, dead by 1260; velocity of falling body "is observed to be continuously multiplied"

Thomas of Bradwardine, about 1290-1326+, velocity as instantaneous rate, led to conclusion that constant forces produce constant accelerations

Nicolas Oresme, before 1377, Speed is proportional to time of fall rather than distance.

Mertonians/Calculatores, succesors to Thomas of Bradwardine

Copernicus b1473-1542?

Maestlin, Kepler's teacher

Tycho Brahe, top astronomer, Kepler became assistant

Thomas Hariot, 1598-1605, distance=time square

Johannes Kepler, 1609, 1619, R-cubed=T-squared; therefore V=1/R

Hans Lippershey Galilea 1609, 1630, V=sqrt D

Marin Mersenne 1626, V=sqrt H(fluid), per Boyle, 1670

Baliani, 1627, letter to Castelli acknowledging Galileo's knowledge of continually augmented acceleration.

Gasparo Berti, d 1639-44, preceeded Torricelli's experiment.

Viviani, Galilea's assistant

Torricelli, etc., 1644 (1608-1647; credited with inventing barometer about 1643

Huygens 1673, centrifugal force = (M) V-squar/D

(Boyle) Halley Newton Hooke

Robison 1767(?), 1802, 1822

Aepinus Franklin Daniel Bernoulli John Canton, 1712-1772 Priestly Tobias Mayer (1760) Lambert (1766) Coulomb Cavendish Dalton Graham Wheatstone Frauenhofer

van't Hoff Abegg Ostwald, Nernst, Kohlrausch, Bousfield Lebon Arrhenius, Bayliss, Harry Jones Fitzgerald-Lorenz Michelson-Morley Einstein, Max Born, etc. Flint Fuller

One of the most interesting aspects of study in the history of science is learning how basic concepts referred to within the context of one or another of a small select group of scientists, e.g. Aristotle, Ptolemy, Copernicus, Newton, etc., may turn out to be the result of the accumulated efforts of successive waves of scientists over generations or even centuries.

UNI-SCIENCE ABSTRACTS

These abstracts comprise an outline for an introductory course and textbook tentatively to be titled

"HYDRATION -- Introduction and Validation of Flint's Description of Hydrational Potentiality".

CONTENTS

[A]

"An Algebraic Approach to Unification." S.H. Shakman

Anticipated by Newton* and sought by Einstein**, a mechanical/algebraic approach to unification of chemistry, physics & biology may be found in Lewis H. Flint's description of hydrational potentiality. Flint's initial discovery (hypothesis), that the sum of atomic number and hydration number for each of the lighter element ions may total 23, was

(1) based on a combination of principles advanced during the 19th Century***: (a) Graham's law of diffusion (gases), (b) Kohlrausch's and van't Hoff's works relating to solute ions and behavior analogous to gases, and (c) Abegg and Bodlander's observation that the hydration potential of solute ions varies inversely with ionic weight; and

(2) derived from information in W.M.Bayliss' PRINCIPLES OF GENERAL PHYSIOLOGY (1915) concerning the K+, Na+ & Li+ ions: (a) Nernst conductivity values of 65.3, 44.4 & 35.5****, and presumably (b) Bousfield's proposed hydration nos. of 4, 8 & 16. Flint used alternate hydration nos. of 4, 12 & 20; multiplied each by 18 (for weight per water molecule); added these values (72, 216 & 360) to atomic weight values of 39, 23 & 7 for K, Na & Li to derive hydrated weight totals (111, 239 & 367); calculated inverse square roots of these totals (.0949, .0647 & .0522) and found these relative values to correlate well with relative Nernst conductivity values.

*NEWTON, I.(1686), in Principia (1687), Preface to 1st Ed. "I am induced by many reasons to suspect that they [the phenomena of Nature] may all depend upon certain forces by wihch the particles of bodies by some causes hitherto unknown, are either mutually impelled towards one another, and cohere in regular figures, or are repelled and recede from one another."
**EINSTEIN, A.(1954), in Relativity (Princeton 1956),App.II. "From the quantum phenomena it appears to follow with certainty that a finite system of energy can be completely described by a finite set of numbers (quantum numbers). This does not seem to be in accordance with a continuum theory, and must lead to an attempt to find a purely algebraic theory for the description of reality."
***FLINT, L., Behavior Patterns of Hydration (1964), 15-30.
****FLINT, L.H., J. Wash. Acad. Sci., 22 (1932), 99.

Presented 17 February 1987 at the 153rd Annual Meeting of the American Association for the Advancement of Science in conjunction with four abstracts - [B],[G],[M], and [N] herein (#110-113 in Book of Abstracts, p.92-3). Proposed for 1988 AAAS Meeting # 0925.16; withdrawn 9/25/87. A partial translation of Abegg and Bodländer, Zeit. f. Anorg. Chem. 454-499 (1899), Euler & Bredig:
                    Abegg and Bodländer, p. 490-91:
  "One of the first proofs of the existence of such hydrated ions is built on considerations of observed mobilities of ions. So are the findings of Euler[1] and even earlier of Bredig[2] the observation that Cl, Br, I

as ions travel equally fast, while the neutral molecules with very different speeds diffuse, very plausibly in the sense would mean, that the various ions are combined with so much water, that the standard mass of Halogens is no longer valid. ... one might for example expect that the due to the increasing atomic weights of lithium, sodium, and potassium, they would exhibit decreasing conductivities; however exactly the opposite occurs; the heaviest atom K, despite its greatest atomic volume, migrates the quickest, lithium the slowest. One assumes that this is due to hydration of ions, whereby lithium must be considerably more strongly-hydrated than sodium, and these moreso than potassium. Such a variation in hydration degree, as follows from the above ... must mean that the lightest ions possess the greatest addition-potential.

" ... the lighter ions H+ and OH-, as concluded fromtheir large mobilities, do not hydrate."

1. Euler, <u>Wied. Ann.</u> (1897) **63**, 273: Even earlier were hydrated ions repeatedly assumed, for example by Ostwald, <u>Lehrb. d. allgem. Chem.</u> (2. Aufl., Leipzig 1893) **2**, 1, 801

2. Bredig, <u>Zeitschr. phys. Chem.</u> (1894) 13, p. 242, also refers to Ostwald, "Bereits Ostwald has repeatedly thereon indicated that the mobility of elemental ions, die nur aus einem Elemente resp. Atome bestehen [existence], is a distinct periodic function of atomic weight."

On p. 262, Bredig anticipates Flint's exploitation of the analogy between gases and solute ions, citing Wüllners and Ostwald:

"Very obviously is here the place which in <u>Wüllners Lehrbuch der Experimentalphysik</u> (4. Aufl. I, 543) about the resistance of air against the motion of Körpers find and which well mutually even the migration of ions would explain ..."

" ... By the hypothesis, that the ionization is a hydration-prosses (here citing Ostwald pp. 798-801), could one perhaps derive the fact thereby to "explain" that for example the electronegative chlorine ion a greater water "hülle" with [p.263] itself carries, than that with less ionization tendency is endowed Iodine, and that in spite of the less atomic weight etc., is not more mobile than is Iodine. ..."

Flint likely first attempted to calculate relative velocity/conductance from Bousfield's numbers, using Graham's law, compared the result with Nernst's conductance values, and then effected a simple adjustment.

If Bousfield's work had indeed so facilitated Flint's initial calculations, as would appear logical, it is interesting that a subsequent edition of Bayliss's book [*11 (1924 (p. 178)] omitted reference to specific hydration numbers:

Bayliss (1924 p. 178) had noted "The methods for determining the degree of hydration of ions are not reliable enough for definite numbers of water molecules to be assigned to different ions. It appears, however, that of all ions hydrogen is least hydrated and among the alkali metals the hydration increases with decreasing atomic weight (Nernst, 1923 p. 446)."

So it would appear that Flint just slipped through the window.

HAWKING: WRONG – BUT CAN TWO WRONGS MAKE A RIGHT?

Hawking [p11-12] refers to the general theory of relativity [for very large phenomena] and quantum mechanics [for the very small phenomena] as "the great intellectual achievements of the first half of [the 20th] century; "however, these two theories are known to be inconsistent with each other – they cannot both be correct."

Thus Hawking's assessments of both as "the great intellectual achievements" is nonsense. Perhaps "great intellectual mind-games would be more appropriate.

Hawking's book's "major theme ... is the search for a new theory that will incorporate them both – a quantum theory of gravity. But if either quantum or relativity is not correct, and Hawking does not specify which, perhaps it is both? In this case would "two negatives make a positive", or would the contrary "two wrongs don't make a right" prevail? Have we had enough of this nonsensical rut that this "genius" and big physics has led us down?

If, as Hawking states "our goal is nothing less than a complete description of the universe we live in", we are in essence there with the integration of Flint's description of hydrational potentiality within its properly exalted place within the factual world of Aristarchian heliocentricity, the magnificent harmonic laws of Kepler and the embodiment of gravitational reality in that supreme chronicler of science, Newton – all algebraic.

As Einstein so aptly stated at the conclusion of his last book on Relativity [Princeton edition, 1955], the inconsistencies between quantum and continuum theories "must lead to an attempt to find a purely algebraic theory for the description of reality." We can only improve on Einstein by deleting the words "theory for the", as indeed Flint's description is not a theory.

Flint's laws are calculated directly from indisputably, incredibly well-documented data involving very small phenomena, with a methodology wholly consistent with the law of kinetic energy and Newton's laws of gravity for very large phenomena. That is your unification, not a theory – an algebraic fact!

[B]

Conductivity of Positive Ions.   S.H. Shakman

A notable order of agreement between calculated and observed
conductivity values (within 10% for a majority of positive
inorganic ions listed by CRC*) may be derived when ions are
assigned [either anhydrous (a) or maximally hydrated (h)] weight
values (Wa,Wh) in accord with Flint's description of hydrational
potentiality**; and relative conductivities are calculated as
inverse square root of weight (in accord with Graham's law of
diffusion, as first treated by Flint***).

Table - Equivalent Conductivity, Calculated vs Observed

Ion:	H+	Li+	Na+	K+	Rb+	Cs+	[H+:
W(a or h)**	4(a)	350(h)	222(h)	94(h)	76(a)	112(a)	K+]
1//W	.5000	.0535	.0671	.1031	.1147	.0945	4.85
Calculated			[BASE]				
Conductivity	373.1	39.9	[50.08]	77.0	85.6	70.5	4.85
Observed	349.65	38.66	50.08	73.48	77.8	77.2	4.76

*CRC Handbook of Chemistry and Physics, 1985-6, D-167-8.
**FLINT, L.H., Behavior Patterns of Hydration (1964), 21:
  Wa=2(atomic no.+/-valence); Wh=Wa+18H, when H=23n-(atomic
  no.+/-valence), H varies from 23 to 0 and n from 1 to 4.
***FLINT, L.H., J. Washington Acad. of Sci., 22 (1932), 98.

©1987 AAAS Abstract #111; (c) 1985 SHShakman Txu219626

M. Berkowitz and W. Wan, *Journal of Chemical Physics,* (1 January 1987, p. 377] "To describe
the limiting ionic mobility on a molecular level is a very challenging task ..."
   Sharma 1984, p. 32, notes that "According to existing theories ... the sodium ion is postulated to
be differentiated from the postasium ion by the biological membranes because its radius (.95A) is
different from that (1.33A) of potassium and also because the two ions have different "carrier"
molecules to transport them across the membrane. ... The existing postulates cannot explain why the
conductance for the larger K+ ion with radius 1.33 A is higher than that for the smaller Na+ ion
(0.95A). Thus, ion size cannot be the sole and sufficient basis of biological descrimination."

   Flint, 1964, p. 18-9: In that his initially-determined hydration numbers were integers as per
[A] above, Flint had next postulated integers for ionic weights (twice the atomic numbers) and
then:
   "The third and final step which marked the definitive diagnosis of hydrational potentiality was
the introduction of the postulate of a specific change in weight with ionization. ... It was a drastic
revolutionary and seemingly incredible step, but it permitted an integration of observational data
with a satisfying convincing nicety. ...
   "At the time of the discovery in 1932, I had attained the age of 39, an age held and maintained
by no less an authority than Jack Benny to represent the very peak of perfection in
a numan male. ..."

The simple manner in which the hydration number for the series "Li+ Na+ K+" appeared to rachet down as atomic number increased, one per one, justifiably spurred Flint's emphasis on relative atomic number in his subsequent works on hydration. However, in order to obtain satisfactory agreement on conductivity calculations for some particularly prominent ions, e.g. H+ and OH-, Flint hypothesized a shift in relative atomic number values equal to one unit per unit of ionization (Z +- C= Z').

Such an hypothesis is not without precedent: The atomic number of some radioactive elements is known to increase with the loss of a nuclear electron;*22.1

Similarly, in 1921 F.W. Aston proposed the concept of a "packing" effect which might allow for the mass of an atomic nucleus to be less than that of its "constituent charges"*24 [Aston 1921, in " , p. 22.in Prout]; and in 1860 Marignac suggested that the weight of a grouping of primordial atoms, in the form of a (more complex) chemical atom, "might not be exactly the sum of the weights of the primordial atoms composing it".*23

[F.W. Aston, ISOTOPES, ARNOLD 1923, P. 101: " In the nuclei of normal atoms the packing of the electrons and protons is so close that the the additive law of mass will not hold and the mass of the nucleus will be less than the sum of the masses of its constituent charges.]

Flint also discussed in detail how the concept of a shift in atomic-number-equivalent values with ionization might allow for a possible reconciliation with conventionally-appraised weight values.*5f2 In the least complicated cases, this involved the possibility that conventional atomic weight values might represent "combining weight" values halfway between respective neutral and ionized states; e.g. the theoretically neutral sodium atom would "weigh" 22, the Na+ ion, 24, and the "combining weight" would be 23. Likewise the hypothetical "combining weight" of Li+ would be 7, that of K+ would be 39, and H- would be 1, all values approximating conventionally-appraised weight values.

In his calculations, Flint doubled all atomic-number-equivalent values [Z, Z', and Z'h; and characterized them as "ionic weights" - neutral (W = 2Z), anhydrous (Wa = 2Z') and hydrated (Wh = 2Z'h). Conventionally-appraised atomic weight values cannot be substituted in these calculations (as in Table 1) without loss of agreement with observations, most notably in the case of H+. Moreover, this would also involve abandoning the first principle of atomic number*EB91 as the basis of calculations. As Speakman asserted in 1947, atomic number is "a more fundamentally important property of an element than its atomic weight"

Elsewhere in our work we shall continue to use the atomic number (Z) and adjusted atomic-number-equivalent values (Z', Z'h) and not otherwise refer to or utilize Flint's method of doubling these values and referring to them as "ionic weights"; this will hopefully help to avoid constant conflict over the question of what properly constitutes "weight". (However, in the below section on interdiffusion, we shall utilize Flint's "ionic weight" values, insofar as the main purpose there is to directly compare Flint's "ionic weight" of hydrogen gas with its conventionally-appraised weight value within calculations of interdiffusion.)

A further measure of validation may be derived from the ability to extend this system to a majority of positive ions, as noted above and illustrated below.

Table B-2 below displays observed conductivities for the 22 CRC-listed mono-atomic positively-charged elemental ions with $Z = 1$-19, 55-80, plus OH-, listed in the CRC Handbook (1967-8), and error values (E) for approximations derived as the relative inverse-square-root of Z'h, under one of two hydrational assumptions - maximum or zero hydration in accord with Flint's methodology. Note that error values for all of these ions are 10% or less.

Calculations for both H+ and OH- involve the assumption of zero hydration.

TABLE B-1   Equivalent Ionic Conductivities

Z	C	Obs.(25°C)[17] $(10^{-4}m^2\ S\ mol^{-1})$	Calculation Error (by Hydrational Assumption) (Hmax)	(Hmax/2)	(Hmax/4)	(Zero)
9	OH –	198				06
1	H +	349.65				-06
3	Li +	38.66	-03			
4	Be ++	45	08			
11	Na +	50.08	\<BASE\> 00			
12	Mg ++	53.0	-02			
13	Al +++	61	03			
19	K +	73.48	-05			
55	Cs +	77.2				09
56	Ba ++	63.6				-08
57	La +++	69.7				02
58	Ce +++	69.8				03
59	Pr +++	69.5				04
60	Nd +++	69.4				04
62	Sm +++	68.5				05
63	Eu +++	67.8				04
64	Gd +++	67.3				04
66	Dy +++	65.6				01
67	Ho +++	66.3				05
68	Er +++	65.9				05
69	Tm +++	65.4				05
70	Yb +++	65.6				06
80	Hg ++	63.6				09

TABLE B-3   Equivalent Ionic Conductivities

Z	C	Obs.(25°C)[17] $(10^{-4}m^2\ S\ mol^{-1})$	Calculation Error (by Hydrational Assumption) (Hmax)	(Hmax/2)	(Hmax/4)	(Zero)
9	F –	55.4		-09		
17	Cl –	76.31		00		
20	Ca ++	59.47	-37			
21	Sc +++	64.7			05	
24	Cr +++	67			06	
25	Mn ++	53.5		08		

#	El	Charge	Wt	A	B	C	D
26	Fe	++	54		07		
26	Fe	+++	68			06	
27	Co	++	55		07		
28	Ni	++	50		-04		
29	Cu	++	53.6		01		
30	Zn	++	52.8		-02		
35	Br	-	78.1			-02	-14
37	Rb	+	77.8				-09
38	Sr	++	59.4	09			
39	Y	+++	62	04			
47	Ag	+	61.9				-19
48	Cd	++	54		19	-03	
53	I	-	76.8				05

Dec. 3, 1985 – an important implication of weight change with ionization: Life itself is a (mathematical) manifestation of the 4th dimension in that dissolved substances within us can become sufficiently negatively ionized through metabolic processes, such as or that is through osmosis, to become subject to appraisal as weightless - - to in fact be weightless. At that point of course, matter is transformed into energy. Such transitions are discussed in depth in Flint's books.

THE ETHER

 A striking feature of the successful utility of Graham's law to calculate ionic mobility, and by the V'ant Hoff analogy between solute and gaseous phenemona in general, is the apparent lack of effect of water on the process except as a medium analogous to the role of a vacuum to gaseous phenomena. This argues strongly for the existence of the "ether" or some other "imponderable" medium filling space otherwise thought to be void. In other words, by reverse argument, if solute water fills a role analogous to a vaccuum with respect to solute ions, then one might speculate the existence of some medium filling a the vacuum, call it "ether", acting in an analogous manner with respect to gases therein.

Matthews, Robert, Science **263**, 612-3, "Inertia: Does Empty Space Put Up the Resistance?", discusses the possibility raised by Haisch, B., etal in 1 Feb. 1994 Physical Review A that inertia, once understood, might be controlled. Is this not other than revival of the concept of the "ether"?

ILLUSTRATION OF AN ADVANTAGE TO USING Z+C (v. ATOMIC WEIGHT) IN CALCULATIONS FROM IONIC CONDUCTANCE

Calculations of hydration numbers from: ionic "weights"; calculated as inverse sq. of conductance; as adjusted by INPUT data:

[Hydration number = {(k/conduc.-sq) - INPUT} / (INPUT data for H20)]

BASE =	INPUT: ATOMIC WEIGHTS			INPUT: ATOMIC NOS. (Z)			INPUT: Z + C (valence)		
	H+	Rb+	La+3	H+	Rb+	La+3	H+	Rb+	La+3
Rb+	-3.62	0	1.45	-1.68	0	0.87	0.27	0	1.13
H+	0	0.17	0.25	0	0.08	0.13	0	-0.01	0.04
Li+	4.19	18.84	24.7	7.87	14.68	18.23	17.73	16.65	21.22
Na+	1.45	10.18	13.67	3.77	7.83	9.94	9.5	8.86	11.58
K+	-0.9	3.15	4.77	0.36	2.25	3.23	2.81	2.51	3.78
Cs+	-6.24	-2.57	-1.11	-3.45	-1.75	-0.87	-1.67	-1.95	-0.8
La+3	-6.31	-1.8	0	-3.18	-1.09	0	-1.07	-1.41	0

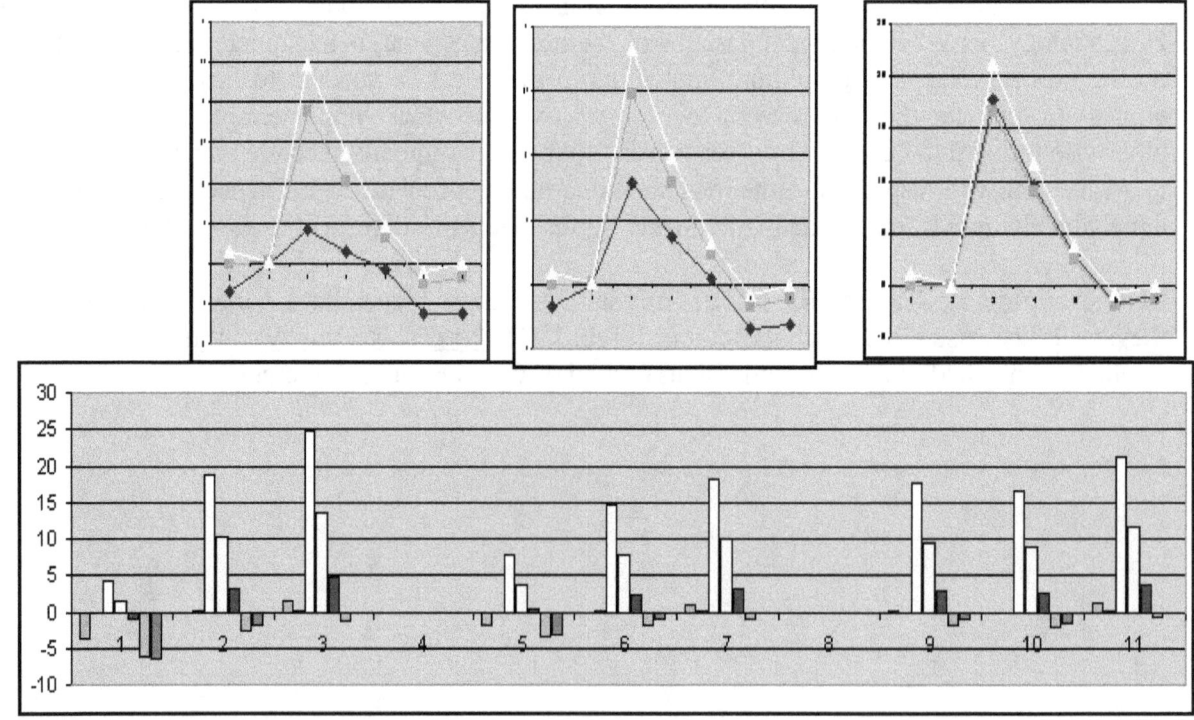

[C]

Flint/Einstein/Newton on Velocity vs Size.   S.H. Shakman

Insofar as velocity of ions in aqueous solution varies directly with ionic conductivity (*1), Flint's use of Graham's law of diffusion in the case of solutes (*2) treated conductivity values as relative (inverse-sq.) measures of ionic weight (*5). Graham's law had preceeded and described a relationship required by the kinetic theory (*3).

Einstein had characterized velocity as varying w/the inverse-sq.-root of mass for particles in colloidal platinum solutions but not for H+ & K+ (*4). For these he calculated displacement as varying w/the sq.-root of conductivity (or velocity); Einstein had also characterized displacement as varying inversely with the square root of diameter as he illustrated in the case of the sugar molecule (*4). Thus velocity might be indirectly characterized as varying inversely with diameter.

Note that both perspectives may be derived from Newton's characterization of resistances as varying (a) "... as the squares of the velocities and the squares of the diameters ..." and (b) "... directly as the squares of the velocities and ...as the quantities of matter ...". (*6)

But as Newton affirmed, "more is in vain when less will serve".(*6)
*1 KOHLRAUSCH, Gottingen Nachrichten, 1876, p. 213.
*2 FLINT, L.H., J. Wash. Acad. Sci., 22, (1932), 99,234.
*3 PAULING, L., Chemical Bond 1967, p. 174.
*4 EINSTEIN, A., Brownian Movement (Dover, 1956), 64,82-85.
*5 SHAKMAN, S.H., 1987 AAAS Abstract #111.
*6 NEWTON, I., Principia (1687), III, Rule 1; II, Prop. 33.

(c) 1987,1990 SHShakman Txu271794. Proposed for 1988 AAAS # 0925.14; withdrawn 9/25/87.

Enstein, *Brownian ...*, p. 12: "The coefficient of diffusion of the suspended substance therefore depends (except for universal constants and absolute temperature) only on the coefficient of viscosity of the liquid and on the size of the suspended particles."

p. 26: "formula for density of radiation corresponding to the frequency v - pv = dd(R/N)(8 pi v-sq./L-cube)T where L is the velocity of light, ... the fact that we obtain in the manner indicated not the true law of radiation but only a limiting law, appears to me to have an explanation in the fundamental incompleteness in our physical conceptions."

On Mistakes/Corrections:
Einstein corrected a 1906 paper in 1911: "Correction of My Paper 'A New Determination of Molecular Dimensions'". And Newton, in Priincipia, asked "... that my labors ... may be examined, not so much with the view to censure, as to remedy their defects."
Einstein, *Brownian ...* p. 114, *Principia*, Preface.

[D]

In search of the elusive hydration number.   S.H. Shakman

1.Average ratio of hydration numbers for Na+:Li+ for 14 sets of data in Amis* (.608)are generally compatible with ratio maximally prescribed by Flint** (11/19=.579) based on atomic no. [Z] & valence[C] per equation H=23-(Z+C) [see fig.1]

Fig.1:	Na+	Li+	Ratio
[FLINT**	11	19	.579]
MacInnes*	2.0	4.7	.426
"	8.4	14	.600
"	14.9	23	.648
"	9.8	14.3	.685
Washburn*	8.4	14	.600
Baborovsky*	8-9	13-14	.630
"	9	14	.643
"	44.5	62	.718
Collet*	5	7	.714
Haase*	13	22	.591
Ulich*	2-4	6-7	.462
Devyatykh*	10	21	.476
Robinson*	5	7	.714
Gapon*	3	5	.600

2.Degree of persistence of hydrates at elevated temperatures, in general agreement w/relative hydration nos. per Flint**, is evidenced by comparing ratios, relative to Na+=1.00, of observed (CRC'67) to calculated conductivities, calculated as inverse-sq.-root of anhydrous(a) [Wa=2(Z+C)] or hydrated(h) [Wh=Wa+18H] weights (for 8 ions with ratios w/in 10% of BASE at 0&C) after Graham per Flint [see fig.2].

fig.2:	(h/a)	0øC	18øC	25øC	50øC	75øC
Na+BASE	(h)	1.00	1.00	1.00	1.00	1.00
K+	(h)	1.01	.97	.95	.91	.89
C2H3O2-	(h)	.98	1.00	1.00	1.02	1.04
Ba++	(a)	.92	.91	.92	.92	.93
La+3	(a)	.99	1.03	1.04	1.07	1.10
SO4--	(a)	1.02	1.01	1.00	.98	.98
C2O4--	(a)	.92	.89	.88	.86	.84
OH-	(a)	1.08	1.06	1.01	.93	.83

*AMIS,E.& J.HINTON,Solvent Effects...(1973) Tables 3-1,4,15.
**FLINT,LH,J.Wash.Acad.Sci.22(1932)97+;-Behavior...(1964)21.

Fig.2. example: calculation for K+=.97 at 18&C: Observed conductivities for K+ and Na+ are 64.6 & 43.5 resp.; anhydrous weights[(Wa)=2(Z+C)] 40 & 24, hydrated weights [(Wh)= Wa+18H] 94 & 222, and calculated conductivities 1//94=.1031 & 1//222= .0671 resp.; (64.6/.1031)/(43.4/.06712)=.97.

Error	Pos (49)				Neg (59)				Total (108)			
	max	1/2	0	T	max	1/2	0	T	max	1/2	0	T
w/in 10%	8	7	17	32	7	9	8	24	15	16	25	56
w/in 20%	19	8	8	35	17	11	11	39	36	19	19	74

[E]

A Lock for Flint: Diameter & Conductivity.   S.H. Shakman

For a majority of eight (hydrated) ions listed*, diameter* &
conductivity** may be approximated as cube-root of volume &
inverse-square-root of weight, resp., of fully-hydrated ions
per Flint***. Anomolous results may be improved (alt.) for
(a) both size and conductivity of Cl- by assuming hydration
number [H] of half the prescribed maximum; (b) conductivity
of SO4-- by assuming ion is anhydrous when measured for conductivity; & (c) diameter of HPO4-- by speculatively deriving it as the cube-root of 2 fully-hydrated ionic volumes.
Assumptions in (a) & (b) above, assigned to Cl- & Na+ ions resp., yield approximations of specific gravities for NaCl solutions comparable to those for 5 other solutes
assumed to be fully hydrated (see [G] "Specific Gravity"; error values for all 6 also describe regular curves).****

IONIC SIZE AND CONDUCTIVITY

Ion	Diameter		Conductivity		
	Obs*	Calc	Ob**	Calc	H
Cl-	.96	1.28	76.3	59.4	7
" alt.	.96	1.04	76.3	76.6	3.5
K+	1.00	1.02	73.5	76.9	3
Na+ BASE	1.47	1.47	50.1	50.1	11
HCO3-	1.65	1.66	44.5	40.0	16
CH3COO-	1.80	1.68	40.9	40.0	16
SO4--	1.84	1.89	80	33.2	23
" alt.			80	77.8	0
H2PO4-	2.04	1.84	33	34.3	21
HPO4--	2.58	1.89	33	33.2	23
" alt.	2.58	2.39			

*GANONG, W. F., Review of Medical Physiology (1975) 12. ** CRC
(1985-6) D167-8. ***FLINT,L.H., Behavior Patterns of
Hydration (1964) 21-30: Wa[anhydrous weight]=2(atomic # +-
valence); Wh[hydrated weight]=Wa+18H; H[hydration#] =23n-(
atomic # +-valence) [H=23 to 0, n=1 to 4]; Vh[hydrated vol-
ume]=Wh/{1+(Wa/Wh)}.   ****SHAKMAN, S.H., AAAS (1987) #113.

AAAS Pacific Div. Proceedings Vol.7(1988), p. 42.

VOLUME AS CUBE OF IONIC RADII

Shankland, Robert S., MacMillan N.Y. 1955, Atomic and Nuclear Physics, p. 290, Neutron scatering
experiments have been made with elements from Be to U distributed throughout the periodic system,
and the nuclear radii so determined are proportional to $A^{1/3}$

Flint then turned his attention to attempts to validate the description using specific gravities, etc.

[F]

OPTIONS:3+3 [Hydration * Specific Gravity]. S.H. Shakman

A. Agreement (w/in 10%) between observed and calculated
conductivity values for majority of (49) positive ions calc. as per
1987 AAAS Abstract #111 (see [B]) may be extended to majority of total 108 ions in CRC 1985, D167, when 3rd hydrational
assumption is allowed (max-per-Flint)/2. Ions in each of 3
below groups often relate sequentially &/or thru Mendeleev.:
  (1)Hydration # [H]=zero for all 12 listed positive
lanthanide ions; H+,OH-; Cs+,Rb+; SO3-2,SO4-2,S2O4-2;
  (2)H=maximum-per-Flint for Li+[H=19],Na+[11],K+[3]; Be++
[17],Mg++[9],Sr++[6]; HPO4-2[23],H2PO4-[21],H2PO2-[14];
  (3)(H=max-per-Flint)/2 for Mn++[H=19/2],Fe++[18/2],Co++
[17/2],Ni++[16/2],Cu++[15/2],Zn++[14/2]; F-[15/2],HF-[5/2],
Cl- [H=7/2],ClO2-[14/2],ClO3-[6/2],BrO3-[11/2],IO3-[16/2].
  B. Flint in 1964 [Behavior Patterns...,Chs.3,15] derived
hydrated volume (Vh) as a function of anhydrous (Wa) and hydrated (Wh) wts. [Vh=Wh/(1+Wa/Wh)]; subtracted Vh in cc from 1000
to derive volume of free water in a 1-liter solution; added
added same free-water amount in grams to wt. of hydrated solute; and divided result by 1000 to calculate specific gravity.
  Curiously Flint used 2 distinct methods to determine ratio
(Wa/Wh) in above equation for Vh:
(1) ratio for total molecule [used in [G] 1987 AAAS Abstract#113; &
(2) average of individual ionic ratios.
As a 3rd option, individual ionic hydrated volumes might be summed.

W.G. Ganong, *Rev. of Med. Physiology* (1975), p. 12]: "The ions in the body are hydrated, and
although the atomic weight of potassium (39) is greater than that of sodium (23) the hydrated
sodium ion, i.e., Na+ with its full complement of water, is larger than the hydrated potassium ion."

Encyclopedia Britannica (198?): "... the ions Cl-, Br-, and I- have almost identical mobilities
although it might be expected that the heavy and bulky I- ion would a much lowermobility than the
lighter and smaller Cl- ion. Even more surprisinigly, the small and light lithium (Li+) ion has only
about half the mobility of the heavy cesium (cs+) ion."

Abegg & Bodlander in 1899[Z.Anorg. Chem.20,491] thought resp. conductivities evidenced
anhydrous H+ & OH- ions; Flint indicated such for H+,Rb+,Cs+ w/Cl- [1932,p.233].

Bousfield in 1906[Phil.Trans.206A,124] posited H=3.5 for Cl-.
See also Bousfield, W.R.., M.A., K.C., "Ion Size in Relation to Molecular Physics, together with a New
Law Relating to the Heats of Formation of Solid, Liquid, and Ionic Molecules", Oct. 18, 1912

[G]

      Specific Gravity of Aqueous Solutions. S.H. Shakman

Specific gravity (S.G.) for a number of solutes may be approximated using the equation: S.G.= ((1000+ [g/l / (1+ Waz/Whz)])) /1000, as derived from Flint*; when g/l= grams/liter anhydrous solute; Waz is the sum of anhydrous ionic weights (Wa), each equal to: 2(atomic no. +/- valence); Whz is the sum of hydrated ionic weights (Wh), each equal to: Wa+18H, when H (hydration no.) = 23n - (atomic no. +/- valence), with H varying from 23 to 0 and n from 1 to 4. For example, UO2(NO3)2 is assumed to consist of U+6((Wa=196, Wh=196 [H = 23(4) - (92+6), thus U+6 does not hydrate])); 2 O--(Wa=12, Wh=318); and 2 NO3-(Wa=60, Wh=348). Thus Waz=340, Whz=1528, Waz/Whz=.2225, and calculated S.G. at 2% solute = (1000+ 20.34/1.2225) /1000 = 1.017. Table below lists, for solutions of "A%" (2-20%) solute, observed and calculated specific gravities and error (obs./calc. - 1) for UO2(NO3)2; and error values only for 4 other solutes.

Table - Specific Gravity of Aqueous Solutions

A%	UO2(NO3)2			MgI2	CdCl2	CaCl2	CoCl2	CdI2	CsBr
	Obs.**	g/l	Calc.	Error = (Observed**/Calculated - 1)					
2	1.017	20.34	1.017	.0000	-.0015	-.0019	-.0010	-.0008	-.0010	-.0016
4	1.036	41.44	1.034	.0019	-.0013	-.0015	-.0005	-.0002	-.0004	-.0015
8	1.072	85.76	1.070	.0019	-.0012	-.0010	-.0004	.0007	.0003	-.0014
12	1.110	133.20	1.109	.0009	-.0010	.0007	-.0013	.0012	.0006	-.0015
16	1.149	183.84	1.150	-.0009	-.0009	.0012	-.0026	.0013	.0010	-.0016
20	1.194	238.80	1.195	-.0008	-.0010	.0014	-.0048	.0005	.0015	-.0018

*FLINT, L.H., Behavior Patterns of Hydration (1964), 21, 30.
**LANGE'S Handbook of Chemistry, 10th Ed. (1967), 1169-82.

                    Specific Gravity
                    Calculations

Sec. 1   Hydration calculations based on ratio between sum of anhydrous ionic weights and sum of hydrated ionic weights
              1. Uranyl Nitrate - UO2(NO3)2; U+6, 2O--, 2NO3-; Wa=340, Wh=1528

		%	Wz	g/l-A	g/l-H	Vh	Vfw=Wfw	Wz-calc	Wz-obs	Error=o/c	Error-c/o
340	1528	1	1008	10.08	45.30071	37.0554	962.9446	1008.245	1008	-0.00024	0.000243
340	1528	2	1017	20.34	91.41035	74.77249	925.2275	1016.638	1017	0.000356	-0.00036
340	1528	4	1036	41.44	186.2362	152.3388	847.6612	1033.897	1036	0.002034	-0.00203
340	1528	6	1054	63.24	284.208	232.4785	767.5215	1051.73	1054	0.002159	-0.00215
340	1528	8	1072	85.76	385.4155	315.265	684.735	1070.151	1072	0.001728	-0.00173
340	1528	10	1091	109.1	490.3082	401.0658	598.9342	1089.242	1091	0.001614	-0.00161
340	1528	12	1110	133.2	598.6165	489.6606	510.3394	1108.956	1110	0.000942	-0.00094
340	1528	14	1129	158.06	710.3402	581.0492	418.9508	1129.291	1129	-0.00026	0.000258

340	1528	16	1149	183.84	826.1986	675.8198	324.1802	1150.379	1149	-0.0012	0.0012
340	1528	18	1171	210.78	947.2701	774.8548	225.1452	1172.415	1171	-0.00121	0.001209
340	1528	20	1194	238.8	1073.195	877.86	122.14	1195.335	1194	-0.00112	0.001118
340	1528	22	1218	267.96	1204.244	985.0559	14.94407	1219.188	1218	-0.00097	0.000975
340	1528	24	1243	298.32	1340.685	1096.663	-96.6632	1244.022	1243	-0.00082	0.000822
340	1528	30	1322	396.6	1782.367	1457.953	-457.953	1324.414	1322	-0.00182	0.001826
340	1528	40	1466	586.4	2635.351	2155.683	-1155.68	1479.668	1466	-0.00924	0.009323
340	1528	50	1649	824.5	3705.4	3030.97	-2030.97	1674.43	1649	-0.01519	0.015422

2. MgI2; Mg++, 2I-; Wa=236, Wh=1010

236	1010	2	1014.9	20.298	86.86856	70.41512	929.5849	1016.453	1014.9	-0.00153	0.001531
236	1010	4	1032.1	41.284	176.6815	143.217	856.783	1033.465	1032.1	-0.00132	0.001322
236	1010	8	1068	85.44	365.6542	296.3971	703.6029	1069.257	1068	-0.00118	0.001177
236	1010	12	1106.5	132.78	568.2534	460.6227	539.3773	1107.631	1106.5	-0.00102	0.001022
236	1010	16	1147.8	183.648	785.9512	637.0872	362.9128	1148.864	1147.8	-0.00093	0.000927
236	1010	20	1192	238.4	1020.271	827.0256	172.9744	1193.246	1192	-0.00104	0.001045
236	1010	25	1251.9	312.975	1339.427	1085.731	-85.7313	1253.696	1251.9	-0.00143	0.001434
236	1010	30	1318	395.4	1692.178	1371.669	-371.669	1320.509	1318	-0.0019	0.001904
236	1010	35	1391.4	486.99	2084.152	1689.401	-689.401	1394.751	1391.4	-0.0024	0.002408
236	1010	40	1473	589.2	2521.576	2043.974	-1043.97	1477.602	1473	-0.00311	0.003124

3. CdCl2; Cd++, 2Cl-; Wa=164, Wh=758

164	758	2	1014.8	20.296	93.80712	77.12126	922.8787	1016.686	1014.8	-0.00185	0.001858
164	758	4	1032.5	41.3	190.8866	156.9328	843.0672	1033.954	1032.5	-0.00141	0.001408
164	758	8	1069.2	85.536	395.3432	325.0219	674.9781	1070.321	1069.2	-0.00105	0.001049
164	758	12	1110.3	133.236	615.8103	506.2735	493.7265	1109.537	1110.3	0.000688	-0.00069
164	758	16	1153.1	184.496	852.7315	701.0526	298.9474	1151.679	1153.1	0.001234	-0.00123
164	758	20	1198.8	239.76	1108.159	911.0461	88.95386	1197.113	1198.8	0.001409	-0.00141
164	758	24	1247.6	299.424	1383.923	1137.759	-137.759	1246.164	1247.6	0.001152	-0.00115
164	758	28	1300.3	364.084	1682.778	1383.456	-383.456	1299.323	1300.3	0.000752	-0.00075
164	758	32	1355.8	433.856	2005.261	1648.577	-648.577	1356.684	1355.8	-0.00065	0.000652
164	758	36	1417.8	510.408	2359.081	1939.461	-939.461	1419.62	1417.8	-0.00128	0.001283
164	758	40	1484.3	593.72	2744.145	2256.032	-1256.03	1488.113	1484.3	-0.00256	0.002569
164	758	44	1556.3	684.772	3164.983	2602.014	-1602.01	1562.969	1556.3	-0.00427	0.004285
164	758	48	1634.5	784.56	3626.198	2981.191	-1981.19	1645.007	1634.5	-0.00639	0.006428
164	758	52	1720.6	894.712	4135.315	3399.749	-2399.75	1735.566	1720.6	-0.00862	0.008698
164	758	56	1813.1	1015.336	4692.833	3858.1	-2858.1	1834.734	1813.1	-0.01179	0.011932

4. CaCl2/ Ca++, 2Cl-; Wa=108, Wh=378

108	378	1	1006.5	10.065	35.2275	27.39917	972.6008	1007.828	1006.5	-0.00132	0.00132
108	378	2	1014.8	20.296	71.036	55.25022	944.7498	1015.786	1014.8	-0.00097	0.000971
108	378	3	1023.2	30.696	107.436	83.56133	916.4387	1023.875	1023.2	-0.00066	0.000659
108	378	4	1031.6	41.264	144.424	112.3298	887.6702	1032.094	1031.6	-0.00048	0.000479
108	378	5	1040.1	52.005	182.0175	141.5692	858.4308	1040.448	1040.1	-0.00033	0.000335
108	378	6	1048.6	62.916	220.206	171.2713	828.7287	1048.935	1048.6	-0.00032	0.000319
108	378	7	1057.2	74.004	259.014	201.4553	798.5447	1057.559	1057.2	-0.00034	0.000339

108	378	8	1065.9	85.272	298.452	232.1293	767.8707	1066.323	1065.9	-0.0004	0.000397
108	378	9	1074.7	96.723	338.5305	263.3015	736.6985	1075.229	1074.7	-0.00049	0.000492
108	378	10	1083.5	108.35	379.225	294.9528	705.0472	1084.272	1083.5	-0.00071	0.000713
108	378	12	1101.4	132.168	462.588	359.7907	640.2093	1102.797	1101.4	-0.00127	0.001269
108	378	14	1119.8	156.772	548.702	426.7682	573.2318	1121.934	1119.8	-0.0019	0.001905
108	378	16	1138.6	182.176	637.616	495.9236	504.0764	1141.692	1138.6	-0.00271	0.002716
108	378	18	1154.9	207.882	727.587	565.901	434.099	1161.686	1154.9	-0.00584	0.005876
108	378	20	1177.5	235.5	824.25	641.0833	358.9167	1183.167	1177.5	-0.00479	0.004812
108	378	22	1197.6	263.472	922.152	717.2293	282.7707	1204.923	1197.6	-0.00608	0.006114
108	378	24	1218	292.32	1023.12	795.76	204.24	1227.36	1218	-0.00763	0.007685
108	378	26	1238.8	322.088	1127.308	876.7951	123.2049	1250.513	1238.8	-0.00937	0.009455
108	378	28	1260	352.8	1234.8	960.4	39.6	1274.4	1260	-0.0113	0.011429
108	378	30	1281.6	384.48	1345.68	1046.64	-46.64	1299.04	1281.6	-0.01343	0.013608
108	378	32	1303.6	417.152	1460.032	1135.58	-135.58	1324.452	1303.6	-0.01574	0.015995
108	378	34	1326	450.84	1577.94	1227.287	-227.287	1350.653	1326	-0.01825	0.018592
108	378	36	1348.8	485.568	1699.488	1321.824	-321.824	1377.664	1348.8	-0.02095	0.0214
108	378	38	1372	521.36	1824.76	1419.258	-419.258	1405.502	1372	-0.02384	0.024419
108	378	40	1395.7	558.28	1953.98	1519.762	-519.762	1434.218	1395.7	-0.02686	0.027597

5. CoCl2 ; Co++, 2Cl-; Wa=122, Wh=680

122	680				0	0					
122	680	2	1016.4	20.328	113.3036	96.0679	903.9321	1017.236	1016.4	-0.00082	0.000822
122	680	4	1034.9	41.396	230.7318	195.633	804.367	1035.099	1034.9	-0.00019	0.000192
122	680	8	1073.6	85.888	478.72	405.8973	594.1027	1072.823	1073.6	0.000724	-0.00072
122	680	12	1114.8	133.776	745.6367	632.2107	367.7893	1113.426	1114.8	0.001234	-0.00123
122	680	16	1158.7	185.392	1033.332	876.1422	123.8578	1157.19	1158.7	0.001305	-0.0013
122	680	20	1205	241	1343.279	1138.94	-138.94	1204.339	1205	0.000549	-0.00055

Cdl2; Cd++, 2l-; Wa=308, Wh=1262

308	1262	2	1015.3	20.306	83.20186	66.87945	933.1205	1016.322	1015.3	-0.00101	0.001007
308	1262	4	1032.8	41.312	169.2719	136.0644	863.9356	1033.207	1032.8	-0.00039	0.000395
308	1262	8	1069	85.52	350.4099	281.667	718.333	1068.743	1069	0.000241	-0.00024
308	1262	12	1107.5	132.9	544.5448	437.7169	562.2831	1106.828	1107.5	0.000607	-0.00061
308	1262	16	1148.9	183.824	753.2009	605.4392	394.5608	1147.762	1148.9	0.000992	-0.00099
308	1262	20	1193.7	238.74	978.2139	786.3095	213.6905	1191.904	1193.7	0.001507	-0.0015
308	1262	25	1254.6	313.65	1285.15	1033.032	-33.0317	1252.119	1254.6	0.001982	-0.00198
308	1262	30	1321.9	396.57	1624.907	1306.135	-306.135	1318.772	1321.9	0.002372	-0.00237
308	1262	35	1396.7	488.845	2002.995	1610.051	-610.051	1392.944	1396.7	0.002696	-0.00269
308	1262	40	1480.1	592.04	2425.826	1949.932	-949.932	1475.895	1480.1	0.002849	-0.00284
308	1262	45	1572.6	707.67	2899.609	2330.768	-1330.77	1568.84	1572.6	0.002396	-0.00239

Specific gravity of NaCl: Calculation options Flint*3

OPTION #1:

In the case of NaCl, specific gravities of aqueous solutions may be approximated as follows (per Flint*3): Assuming zero hydration for Na+ (as in a. above) and 50%-max.hydration for Cl-(H=3.5, as in c. above); anhydrous ionic weight [Wa=2(Z+C)]=56, hydrated ionic weight

[Wh=Wa+18H]=119, & hydrated ionic volume [Vh=Wh/(1+(Wa/Wh))]=80.92; thus 130.28g of anhydrous solute would have hydrated ionic wt. of 276.8450g and hydrated volume of 188.2546cc. This plus 811.7454cc of free water would comprise a one-liter (1000cc) solution. Add the weight of free water (811.7454g.) to that of hydrated solute (276.8450g) for a calculated weight of 1088.7g, v.s. an observed value of 1087.6g (1000 x specific gravity) [Flint*3.]

OPTION # 2
In the case of NaCl, Flint (Behavior Patterns, p. 134, Table 36) assumed the Na+ ion hydrates fully with the resultant Wa/Wh*** {or Z'/Z'h} ratio of .1081081; the Cl- ion remained anhydrous with a Wa/Wh*** {Z'/Z'h} = 1, and the average, .554054 used in calculations, yielding a calculated liter weight of 1.083835 vs 1085.7 (Density, or 1087.6 Specific Gravity)

OPTION # 3
For example in the case of NaCl, assuming H=11 for Na+ and H=0 for Cl-, Wa/Wh*** for Na+=24/222=.1081081 and Wa/Wh*** for Cl-=1; at 12% gpl=130.28g, Na+ - 55.83428g Na+ x Wh/Wa*** = (516.46453g Na+/1.1081081)=466.07774cc + (74.44572g Cl- /2)=37.22281cc Cl- 503.30055 -1000-(74.44572-516.46453 == 1087.6212 v.s. 1087.6g

A%	UO2(NO3)2	MgI2	CoCl2	CdI2	CsBr	NaCl		
	Obs.	Error=(Observed**/Calculated - 1)				(1)	(2)	(3)
2	1.017	.0000	-.0015	-.0008	-.0010	-.0016	.0005	.0013
4	1.036	.0019	-.0013	-.0002	-.0004	-.0015	.0007	.0021
8	1.072	.0019	-.0012	.0007	.0003	-.0014	.0004	.0034
12	1.110	.0009	-.0010	.0012	.0006	-.0015	-.0009	.0037
16	1.149	-.0009	-.0009	.0013	.0010	-.0016	-.0029	.0032
20	1.194	-.0008	-.0010	.0005	.0015	-.0018	-.0054	.0022

Formulas:

Specific Gravity:
  Wz-calc = 1000 + (g/l)(1-1/[1+W/Wh])/(Wa/Wh)

Freezing Point Depression:
 [[((([(g/l-A)/Wa] x nz x 1.86)) / ((Wz - [(g/l-H) - (g/l-A)]))]] x 1000
  = 51.666 (g/l-A) / [Wz + (g/l-A) - (g/l-H)]
  = [18600 nz (g/l-A)] / [100 WaWz + (g/l-A)], or in terms of A%: = (18.6 A% nz) / [Wa+.01A%(Wa-Wh)]

Freezing Point Depression Calculations - CaCl2 from 1 to 40%

		%	Wz	g/l-A	g/l-H		Freezing Point Dpression		
		CaCl2/ Ca++, 2Cl-; Wa=108, Wh=378							
						obs	raw	obs	relative
108	378	1	1006.5	10.065	35.2275	0.44	0.52991453	0.44	0.44
108	378	2	1014.8	20.296	71.036	0.88	1.0877193	0.88	0.903158
108	378	3	1023.2	30.696	107.436	1.33	1.67567568	1.33	1.391351
108	378	4	1031.6	41.264	144.424	1.815	2.2962963	1.815	1.906667
108	378	5	1040.1	52.005	182.0175	2.345	2.95238095	2.345	2.451429
108	378	6	1048.6	62.916	220.206	2.93	3.64705882	2.93	3.028235
108	378	7	1057.2	74.004	259.014	3.573	4.38383838	3.573	3.64
108	378	8	1065.9	85.272	298.452	4.275	5.16666667	4.275	4.29
108	378	9	1074.7	96.723	338.5305	5.04	6	5.04	4.981935
108	378	10	1083.5	108.35	379.225	5.86	6.88888889	5.86	5.72
108	378	12	1101.4	132.168	462.588	7.7	8.85714286	7.7	7.354286
108	378	14	1119.8	156.772	548.702	9.83	11.1282051	9.83	9.24
108	378	16	1138.6	182.176	637.616	12.28	13.7777778	12.28	11.44
108	378	18	1154.9	207.882	727.587	15.11	16.9090909	15.11	14.04
108	378	20	1177.5	235.5	824.25	18.3	20.6666667	18.3	17.16
108	378	22	1197.6	263.472	922.152	21.7	25.2592593	21.7	20.97333
108	378	24	1218	292.32	1023.12	25.3	31	25.3	25.74
108	378	26	1238.8	322.088	1127.308	29.7	38.3809524	29.7	31.86857
108	378	28	1260	352.8	1234.8	34.7	48.2222222	34.7	40.04
108	378	30	1281.6	384.48	1345.68	41	62	41	51.48
108	378	32	1303.6	417.152	1460.032	49.7	82.6666667	49.7	68.64
108	378	34	1326	450.84	1577.94				
108	378	36	1348.8	485.568	1699.488				
108	378	38	1372	521.36	1824.76				
108	378	40	1395.7	558.28	1953.98				

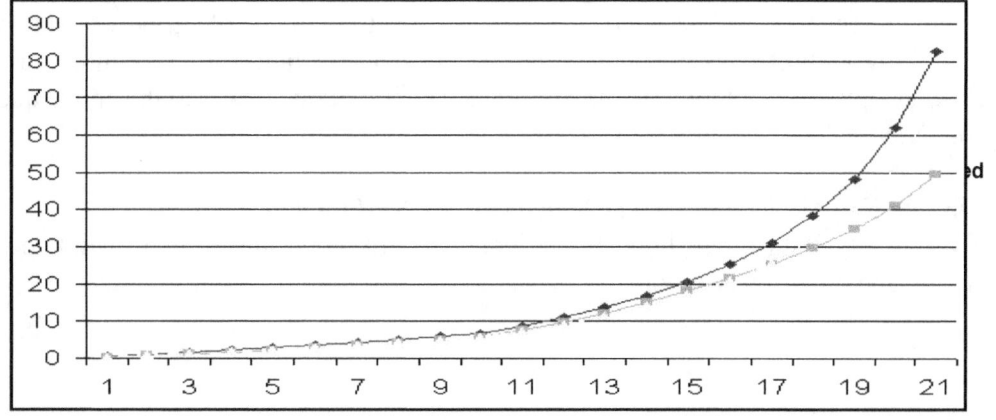

[H]

Organic Compounds - Osmolality vs Volume. S.H. Shakman
Relative observed maximum osmolality (OS) for a majority of
organic compounds listed by CRC* may be approximated (within
10%) as relative theoretical increase in solute volume (DV)
attributable to full hydration of assumed component ions,
calculated in accord with Flint's description, volumetric
definition and equation DV=(gpl/k)((Vhz/2Vaz)-1)**, when
gpl=grams-per-liter anhydrous solute, k=520, and Vhz & Vaz
resp. = sums of hydrated and anhydrous ionic volumes (Vh,Va).

MAXIMUM OSMOLALITY vs VOLUMETRIC INCREASE,ORGANIC COMPOUNDS

Compound	Assumed Ions	gpl*	OS*	DV	OS/DV
ACETIC ACID	CH3-,CO++,OH-	375.6	7.20	6.85	1.05
ACETONE	CO++,2CH3-	98.5	1.767	1.80	.98
ETHANOL	CH3-,CH2++,H+,O--	593.2	26.62	24.97	1.07
FORMIC ACID	CO++,H-,OH-	733.2	23.13	22.46	1.03
GLUCOSE	3CHOH++,3CHOH--	337.4	2.577	2.28	1.13
GLYCEROL	2CH2++,CHOH--,2OH-	439.4	8.33	9.10	.92
LACTOSE	C6H10O5,6CHOH++--	83.5	.266	.26	1.02
METHANOL	C+4,3H-,OH-	595.9	51.8	52.50	.98
OXALIC ACID	CO++,CO--;OH-,OH+	40.7	.587	.64	.92
2-PROPANOL	2CH3,H+,COH-3	231.1	5.87	6.40	.92
CCL3COOH	C+4,OH+,CO--,3Cl-	172.9	2.051	1.95	1.05
UREA	CO++,2NH2-	494.5	9.47	9.06	1.05

*CRC HANDBOOK,1985-6,D221-D269. **FLINT,L.H., Behavior Pat- terns
of Hydration(1964): Wa[anhydrous weight]=2(Z[atomic #]
+C[valence]); Wh[hydrated weight]=Wa+18H; H[hydration #]=23n -(Z+C)
[H=23 to 0, n=1 to 4]; Vh=Wh/(1+(Wa/Wh)); Va=Wa/2.

Supercedes 1987 AAAS Pac.Div.Abstract,"Max.Osmolality...". Proposed for 1988 AAAS Meeting, AAAS # 0925.15;
withdrawn 25 Sept. 1987.

R. PORRET (1816) "demonstrated the existence of a power not before noticed in the voltaic current,
mainly that of conveying fluids (water) through minute pores, not otherwise pervious to them from
positive to negatively-charged sides], and of overcoming the force of gravity. Is not this
electro-filtration, jointly with electro-chemical action, in constant operation in the minute vessels and
pores of the animal system? ... I cannot help thinking that an affirmative answer to the above question is
capable of a good defence."

RESIDENCE OF OSMOLALITY IN THE DOMAIN OF PHYSIOLOGY
Correlations of osmolality with physiology of nutrients and other substances may facilitate deriving
beneficial nutritional and other interventions.

GILBERT LING – FICTION: ION PUMPS AND CHANNELS; ION SELECTIVITY

Gilbert N. Ling *IN SEARCH OF THE CHEMICAL BASIS OF LIFE*
   p. xxvi: ...conclusion that the widely accepted membrane pump theory of the living cell is incorrect."
   p. 122 "There is not enough energy to operate the Na+ pump".
      p. 572   "...present model, wherein the contractile force depends on an osmotic gradient... How precisely this water activity reduction affects contraction is not known and may prove to be a fruitful area for future research."
   p. 582 "How is [the] potential energy trapped in the resting muscle made to perform mechanical work? A rough estimate shows that the theoretically calculated osmotic force may be adequate. ...Suffice it to recognize that the anticipated osmotic force is quite adequate to meet the maximum need demanded by the force of contracting muscle."
   p. 584 "Finally, it is postulated that localized changes in the osmotic activity of water (concimitant with the release of K+ ions in one area and depolarization of water in another) provide the major force for muscle contraction."

Ling, Gilbert N., *A Revolution in the Physiology of the Living Cell*, Krieger, Malabar Fla., 1992
   p. 10 "The sodium pump theory has been without a[n explanatory] mechanism [of action] since its inauguration" ; as per Glynn and Karlish(1975), no hypothesis exists which would account for the working of the "sodium pump".
   p. 28 "in the light of excessive energy needs ... and the fact that cell membranes with finite dimensions cannot accommodate an infinite number of pumps, the disproof of the membrane-pump theory can only be regarded as complete."

WHEN IS SIZE NOT REALLY SIZE; WRONGING THE RIGHT

Sharma 1984, p. 32, notes that "According to existing theories ... the sodium ion is postulated to be differentiated from the postassium ion by the biological membranes because its radius (.95A) is different from that (1.33A) of potassium and also because the two ions have different "carrier" molecules to transport them across the membrane. ... The existing postulates cannot explain why the conductance for the larger K+ ion with radius 1.33 A is higher than that for the smaller Na+ ion (0.95A). Thus, ion size cannot be the sole and sufficient basis of biological descrimination."

ON THE ASCESSION OF SAP IN TREES

Nature **378**, 14 Dec. 1995, 663-4, Ernst Steudle, "Trees under tension" notes that "Because of the difficulty of proving the existence of high tensions, the cohesion theory has occasionally been questioned and osmotic mechanisms have been proposed as an alternative explanation." In this same issue, Pockman, W.T., etal, p. 715-6, "Sustained and significant negative water pressure in xylem", provide indirect evidence supporting the cohesion theory.

Flint's work demonstrates the efficacy of the osmotic mechanisms, in accord with Ling.

[I]

Osmosis: Mechanics/Physiology.   S.H. Shakman

1.Observed max.osmolality* vs calculated volume increase**[DV] yields agreement comparable to free energies of hydration per Kang model***[YK], for the 8 compounds in both Kang and CRC.

2.Result supports concept of osmolality as active mechanical force to be precisely understood and (eventually) manipulated.

MODELS OF SOLUTE BEHAVIOR

Compound	Error: YK	DV
Acetic Acid	-.07	.05
Acetone	-.09	-.02
Ethanol	-.10	.07
Ethylene Glycol	-.20	.08
Glycerol	-.34	-.08
Methanol	-.16	-.05
1-Propanol	-.06	.18
2-Propanol	.00	-.08

3.This would appear to relate to metabolic processes e.g.:
-- Ling:"it is postulated that localized changes in the osmotic activity of water (concomitant with the release of K+ ions in one area and depolarization of water in another) [may]provide the major force for muscle contraction"****;
--Diamond:"evidence is now compelling [that] osmotic force developed by active solute transport [may comprise] the mechanism by which water transport is coupled to [the] active solute transport".****

*CRC HANDBOOK, 1985-6, p. D221+.
**FLINT,L,1964:
  DV=(gpl/520)((Vhz/2Vaz)-1);
  Wa[anhydrous wt.]=2(Z[atomic #] +C[valence]); Wh=Wa+18H;
  H[hydration #]=23n-(Z+C) [H=23 to 0, n=1 to 4];
  Vh=Wh/(1+(Wa/Wh)); Va=Wa/2.]
***Kang,Y.K.,etal.,J.PHYS.CHEM.91(1987)4114.[NIA(AG-00322)]
****LING,G.N,In Search of Physical Basis of Life(1984)584.
*****Diamond,J., in Alfred Benzon Symposium 15(1981), p.355.

Encyclopedia Britannica, 1985(?, under blood?), p. 679, Table:
Concentrations of Potassium and Sodium Ions in Plasma and Red Cells of Certain Vertebrates (milli-equiv. per 1000 ml):

	K+ red cells	plasma	Na+ red cells	plasma
Man	95	4	19	136
Dog	8	4	97	143
Chicken	119	6	18	154

CHICKEN SOUP AND THE POTASSIUM-SODIUM BALANCE

Szuromi, P, Science 263 (11 March 1994), 1353, "Cost at the pump":
  "For animal cells, the potassium concentration inside the cell is higher than it is outside, while the reverse is true for sodium." [As regards red blood cells, while this statement holds true for humans and even moreso for chickens, the reverse is true for dogs. Might therein lie some explanation for the reported therapeutic action of chicken soup for humans?]

[J]

Inorganic Osmolality-Patterns of Divergence.   S.H. Shakman

Observed maximum osmolality (OS)* is compared with maximum volume increase (DV) per equation (after Flint) DV=(gpl/k) (Vhz/2Vaz)-1), when gpl=grams-per-liter anhydrous solute, k=520, and Vhz & Vaz=resp. sums of hydrated (Vh) & anhydrous (Va) ionic volumes. Note patterns for divalent positive ions with SO4--; monovalent positive ions with Cl-; ratios between values listed for K+ & Na+ with monovalent negative ions; and ratios between values for for Cl- & SO4-- with monovalent positive ions.

OS/DV VALUES FOR INORGANIC COMPOUNDS

ION:	SO4--	Cl-	NO3-	OH-	Br-	I-	Cl:SO4
Cs+		3.10					SO4
Li+		3.61					
K+	2.86	7.38	2.79	3.78	5.13	4.06	2.6
Na+	1.40	4.36	1.90	2.70	2.87		3.1
NH4+	1.53	4.01					2.6
H+	3.37	3.46	2.14				1.0
Ag+		1.27					
Pb++		1.21					
Cd++	.95	1.30					1.4
Mg++	1.39	4.38					3.2
Ba++		4.13					
Ca++		19.36					
Co++		3.26					
Sr++		7.46					
Cu++	.84						
Mn++	1.01						
Ni++	.94						
Zn++	.95						
K:Na	2.0		1.7	1.5	1.4		1.8

*CRC HANDBOOK,1985,D221-. **FLINT,L.H., Behavior Patterns of Hydration(1964): Wa=2(Z[atomic #]+C[valence]); Wh=Wa+18H; H= 23n-(Z+C) [H=23 to 0, n=1 to 4]; Vh=Wh/(1+(Wa/Wh));Va=Wa/2.

Flint has suggested tht K+ may act as a catalyst for chemical processes involved in metabolism.

[K]

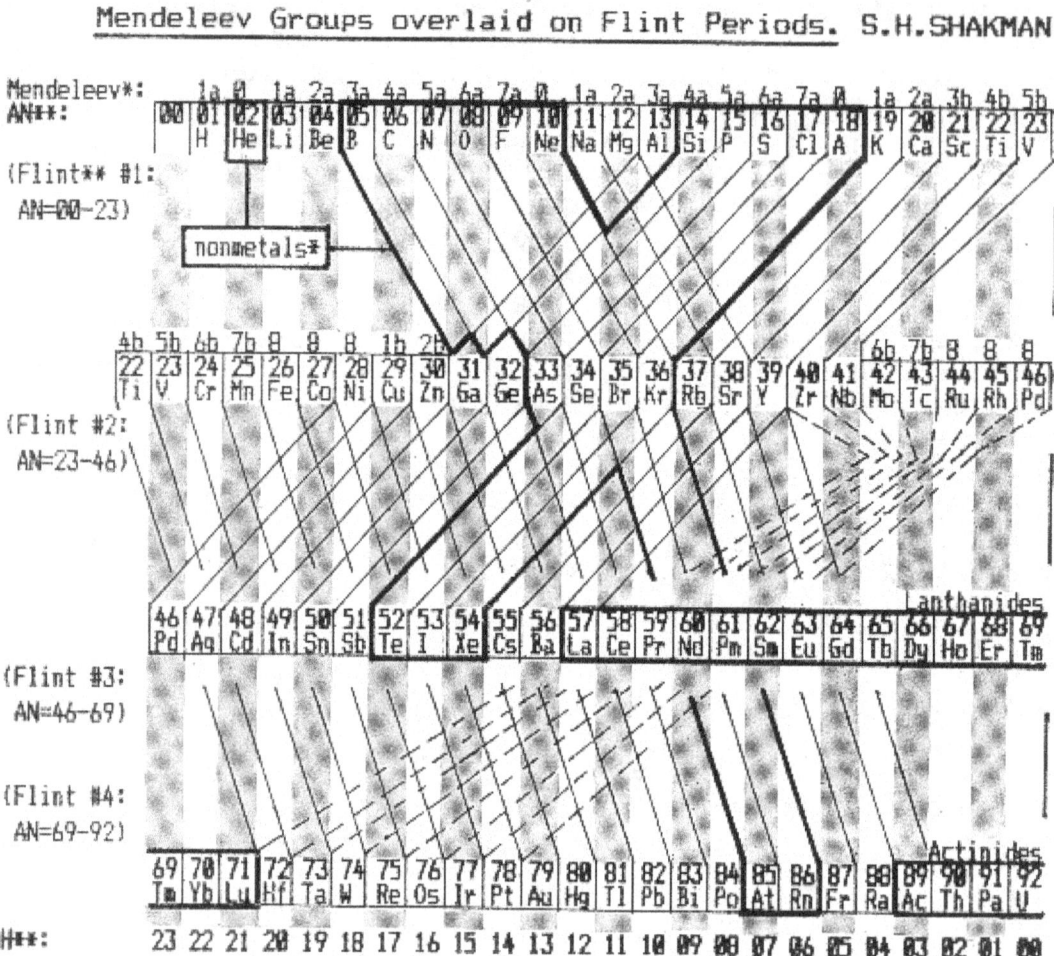

Mendeleev Groups overlaid on Flint Periods. S.H.SHAKMAN

*MORTIMER,CE,Chemistry(1975). **FLINT,LH,Behavior Patterns of Hydration(1964), 21:
H=23n-(AN+-C) [H=(Max.)Hydration No.; n=Period (#1-4); AN=Atomic No.; C=valence].

Published in Proceedings, Pacific Division, American Association for the Advancement of Science, Vol. 6, Part 1, p.39, 16 June 1987, at the 68th Annual Meeting of the Pacific Division American Association for the Advancement of Science. (See [L] for discussion.)

[L] A combined periodic table of elements. S.H. Shakman

"Mendeleev Groups overlaid on Flint Periods" (see [K]) constitutes a combined periodic table of elements which preserves essential information of both taxonomies and may facilitate approaching scientific information from the (alternate or complementary) perspective of either. The value of preserving Mendeleev's periodicity may be inferred from its position in contemporary science; the value of Flint's periodicity may be see in its utility [see [D] &[G].

The utility of Flint's taxonomy serves to emphasize the significance of atomic number as a [quote] more fundamentally important property of an element than its atomic weight [J.C. Speakman, 1947]. Here, atomic number (rather than atomic weight) is utilized as a measure of relative weight. Such a characterization most notably diverges from contemporary thinking in the case of hydrogen [addressed by [N] & [O]].

This combined table of elements (a) may be seen to relate to origins of Mendeleev's periodicity in its symmetrical embodiment of the non-metals and the otherwise somewhat anomolous lanthanides & actinides; (b) may be viewed as geometrically harmonious with the work of R.B. Fuller (see [M]); and (c) constitutes a proposed extension/merger of conventionally-appraised periodicities into an encompassing algebraically-structured system which underlies the fabric of the ponderable universe. The existence of such a system was anticipated by both I.Newton** & A.Einstein*** (see[A]).

*SHAKMAN,S.H., Proceedings,Pac.Div.AAAS,Vol.6,Part 1,p.39.
**NEWTON,Principia(1687).
***EINSTEIN,Relativity(1956)

(c)1987 SHShakman

[Presented on 16 June 1987 at the 68th Annual Meeting of the Pacific Division American Association for the Advancement of Science in conjunction with printing of abstract Mendeleev Groups overlaid on Flint Periods (a combined periodic table of elements) in Proceedings,Pac.Div.AAAS,Vol.6,Part 1,p.39]

The previous year, May 27, 1986, at the previous AAAS meeting, man whose name I did not get, or would cite, came up to my exhibit of a draft version of this chart and told me of Buckminster Fuller's having said there are 184 elements. I said the second 92 would have to be inverse, due to symmetry, etc. He said that Buckminster Fuller had said they were like a glove inside out.

HELICAL PATTERN OF HYDRATIONAL PERIODICITY (after Flint, 1932, 1964, 1968)

As Z+C (atomic number + valence) varies from zero to 23,
"   "   "   "   23 to 46
"   "   "   "   46 to 69
"   "   "   "   69 to 92
   H (hydration number (maximum)) varies from 23 to zero.

Z
+
C   As the Atomic Number + Valence (Z+C) increases: 0-23, 23-46, 46-69, 69-92

$\rightarrow$ $\rightarrow$ $\rightarrow$

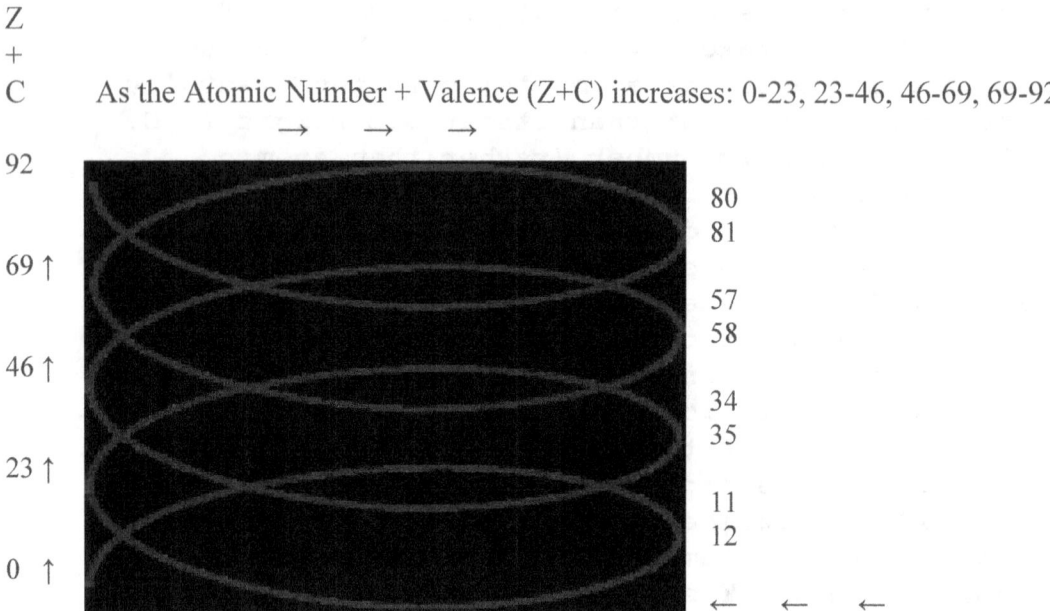

92

69 ↑

46 ↑

23 ↑

0 ↑

80
81

57
58

34
35

11
12

$\leftarrow$ $\leftarrow$ $\leftarrow$

the Hydration Number (H) decreases from 23 to 0, repetitively.

$\rightarrow$ $\rightarrow$ $\rightarrow$
        21 20 19 18 17 16 15 14
    22   13
H: ↑ 23   12 ↓
   0   11
     1   10
       2 3 4 5 6 7 8 9
        $\leftarrow$ $\leftarrow$ $\leftarrow$

Huizenga, J.R., JCE **70**, Sept. 1993, 2pp "Size of the Periodic Table", 3 refs, seeks to answer the "philosophical question" s to "Why is it that our universe contains approximately one hundred elements rather than, for example, ten ... or one thousand...?" The author concludes based on "elementary calculations"involving the approximate A/Z ratio of 5/2, derive a Z_L value of 110, concluding that "size is a direct result of the relative magnitudes of the nuclear and electromagnetic forces."

SOLAR ENERGY OPTICS Afred Jensen, 1975

in *THE SCIENCES* May 1988] New York Academy of Sciences –
"The Painted Equation – An Artist's Rendering of Nature's Laws, by Marcia Tucker

"Jensen was fascinated by the rules that govern behavior of phenomena. Indeed his im waaas to reestablish "man's lost ties with the universal laws of nature."

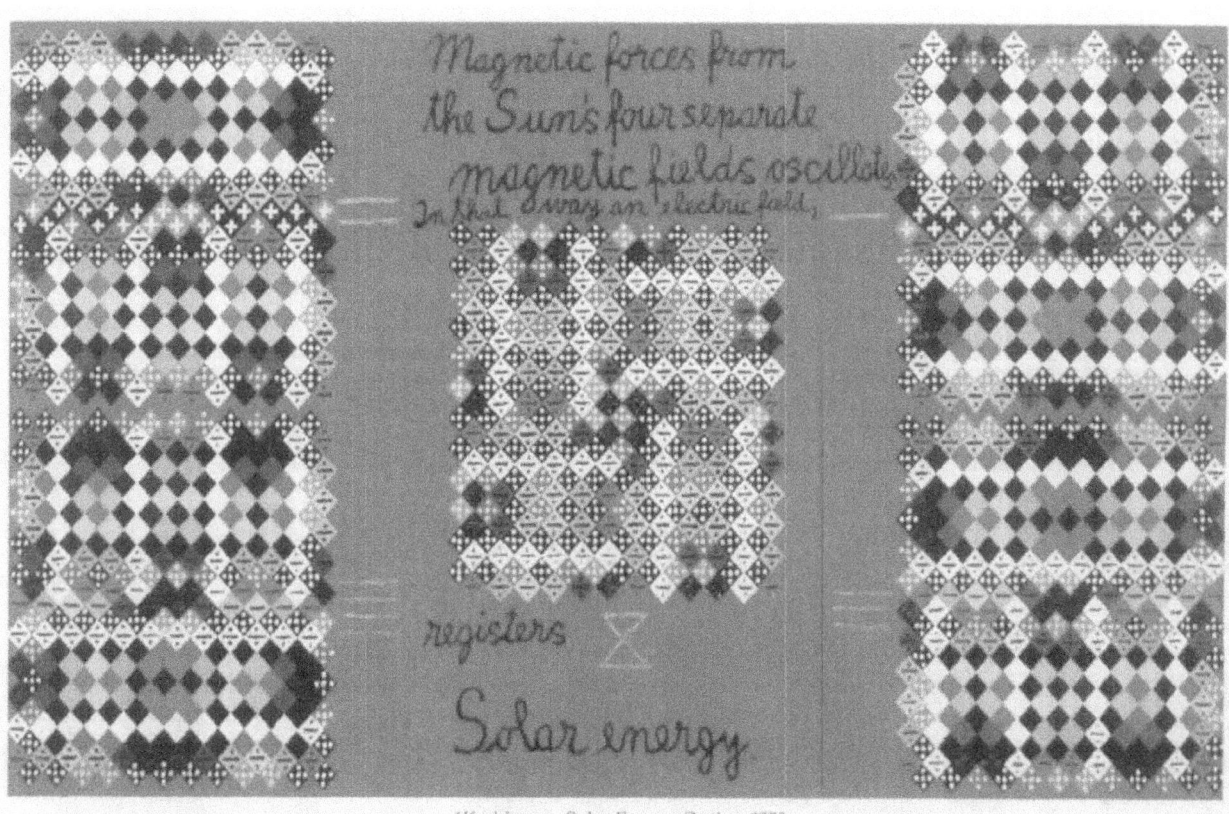

Alfred Jensen, Solar Energy Optics, 1975

THE PAINTED EQUATION

An Artist's Rendering of Nature's Laws

by MARCIA TUCKER

ALFRED JENSEN seemed to have the whole universe in mind when he painted. Talking, in the 1970s, about the inspiration for his work, he would mention solar cycles, Mayan calendars, and Egyptian pyramids; Goethe's color theories and Pythagorean mathematics; the science and technology of ancient China; magnetism, optics, and space travel. Few ideas or world views escaped his attention.

Before the late 1950s, the influence of these far-ranging interests was restricted to drawings on paper. In contrast with his paintings, which were made up of dark, swirling, abstract forms, the drawings were diagrammatical; they consisted of simple geometric shapes and handwritten notes on the nature of color and light. Because the drawings were exploratory, Jensen did not present them to the public, preferring to hang them on the walls of his Tenth Street studio, in Manhattan. Then, in 1957, he turned away from his earlier abstract work and took as the subject of his canvases the brightly colored patterns that, until then, had been the focus of his private research. Jensen became, as he put it, a painter of diagrams.

The change was well suited to Jensen's expanding interest in science. Strictly speaking, a diagram is a graphic design that explains rather than represents; instead of depicting particular objects, it illustrates patterns and principles—exactly the things that concern science. In the paintings that treated, say, organic molecules, Jensen showed not enzymes or hormones but their configurations. That is why these works resemble the stacks of letters and numbers biologists use to illuminate amino acid sequences. Like his scientific counterparts, Jensen was fascinated by the rules that govern the behavior of phenomena. Indeed, his aim was to reestablish "man's lost ties with the universal laws of nature," an ambition that culminated in a series of paintings completed in 1975, six years before he died. One of these climactic works is *Solar Energy Optics*.

JENSEN DABBLED in a variety of scientific subjects, but physics was his lifelong favorite. The work of the nineteenth-century British scientists Michael Faraday and James Clerk Maxwell, and of the early-twentieth century German mathematician Hermann Minkowski, appealed to him particularly. Faraday's chief contribution to scientific thought was field theory, the idea that magnetism is a property not of magnets but of the electromagnetic medium that surrounds them. He believed that not only electricity and magnetism but light and heat were different aspects of the same phenomenon, though he failed to convince most of his contemporaries. That burden fell on the shoulders of Maxwell, whose equations proved Faraday correct. Minkowski, for his part, translated Einstein's theory of relativity into mathematical form, arriving at the concept of a four-dimensional manifold in which space and time are manifestations of a single continuum.

What fascinated Jensen about these three thinkers were their attempts to unify seemingly dissimilar elements by means of increasingly universal symbols—symbols that, in their purest form, were mathematical. In their theories, he also perceived a parallel with ancient and mystical number systems, another of Jensen's enduring interests. Just as physics has come to rely almost exclusively on mathematics as a means of finding order, Jensen's diagrams were always based on numerical concepts.

This is the case with *Solar Energy Optics*. Both the title and the inscription point to what the artist had in mind: the oscillation of the sun's electromagnetic field, including the visible portion of the spectrum, and its relationship to other fields. But the painting no more represents the fields to which it alludes than Maxwell's equations do. As in physics, representation is subordinated to something more fundamental: the patterns that underlie the phenomenon. Thus, the polar nature of electromagnetism is presented in terms of binary signs: plus and minus, or proton and electron; oscillation is illustrated by the multiplication of a diamond shape; and the symmetry of electromagnetic fields is suggested by the overall similarities between the scintillating checkerboards. Though the painting does not include actual numbers, as many of Jensen's other works do, these ordering principles—binary opposition, multiplication, symmetry—are the building blocks of mathematics, indeed, of the world.

In its intent, *Solar Energy Optics* is analogous to the conceptual revolution that occurred at the onset of modern science, when naturalistic description was displaced by Newton's laws of motion. Three, essentially mathematical principles that apply to everything in the universe larger than an elementary particle, Newton's laws represented an enormous leap in the way we understand the world. Every shift in scientific thinking since then —Maxwell's mathematical rendering of electromagnetism; relativity; quantum mechanics—has extended Newton's effort to reduce nature's variety to a handful of simple laws. Jensen's diagrams are the visual equivalent of that effort. Like the theory of everything—the single mathematical formulation that physicists hope will ultimately explain all known forces and elementary particles—this painting is an attempt to bring under one aesthetic roof as much of the universe as possible. *Solar Energy Optics*, in other words, aspires to be a grand unifying painting. ●

MARCIA TUCKER *is director of the New Museum of Contemporary Art, in Manhattan.*

[M]

Fuller-Flint-Einstein Correlations.   S.H. Shakman

Fuller's* reference to "92" as both the number of "self-regenerative chemical elements" and the number of spheres in the third layer of closest-packed spheres dovetails nicely with Flint's** emphasis on the system of 92 naturally-occurring elements.   Particularly striking is Fuller's discussion of the "four of the 24-ness of the Duo-tet Cube" in its uncanny resemblance to Flint's matrix of four overlapping hydrational periods with 24 digits in each (0-23, 23-46, 46-69, 69-92). Fuller: "24 x 4 = 96. But the number of the self-regenerative chemical elements is 92. What is missing ... is the disappearing octa set. ... We have three expendable interior octa and one expendable exterior octa." Presumably Fuller's "three expendable interior octa" may be correlated with the 23, 46 and 69 of Flint's seemingly helical configuration, while the "one expendable exterior octa" may be correlated with Flint's "0". Such correlations may prompt speculation as to a possible correlation between numbers of spheres in the first layer (12) and second layer (42) of closest-packed spheres and Einstein's*** calculated "Z1" values for the "pure gravitational field" (12) and presumed "non-symmetric field" (42). In this context the latter might be subject to reappraisal as symmetrical.

* FULLER, R.B., Synergetics 2 (1975), 418.01, 1033.703-.71.
** FLINT, L.H., J. Wash. Acad. Sci., 22, (1932), 97-119 and 211-217; and Behavior Patterns of Hydration (1964), 21.
*** EINSTEIN, A., Meaning of Relativity, 5th Ed. (A.E., 1954), Princeton U. (1956), 160-1.

[1987 AAAS Abstract #112 (c)1987 AAAS;(c)1987 Txu271794 SHShakman

RIDDLES OF THE NUMBERS – SEEMINGLY ANTICIPATING THE TONE OF BUCKY FULLER, ETC. ON NUMBERS, JOHANNES KEPLER 1596:

Johannes Kepler, MYSTERIUM COSMOGRAPHICUM [THE SECRET OF THE UNIVERSE], 1596,1621, ARABIS BOOKS, NY, 1981: "CHAPTER 10: (1)On the Origin of the Noble Numbers" ...

"Let us now look at the arithmetic of the astronomers, and their sacred numbers, 6, 12, and 60. Now except for the quarter and sixth, that is 15 and 10, all the aloquot parts of sixty are found in these five solids [cube, pyramid, dodecahedron, icosahedron, and icosahedron]. (2) Conversely, except for the plane angles of the octahedron and cube, of which each has 24, everything else, which is countable, is a factor of 60. ... ..."sums of all the edges and angles...the angles of the defining bases will come to 18 ... ."

THE WEIGHT OF THE HYDROGEN ATOM; THE WEIGHT OF H2 GAS

A particularly striking implication of Flint's work concerns hydrogen. Flint's methodology would prescribe an "ionic weight" (W) of 4 for H2 relative to 32 for O2, v.s. conventionally-appraised atomic weight values of approximately 2 and 32 respectively.

For a perspective on this question, Flint turned to the wealth of published data on gas interdiffusion, which he proposed to define for a given gas-pair as the (relative) quotient of (a) the difference in mobilities for each of the two gases, each calculated as inverse-square-root of respective W values, and (b) the difference between these W values.

As Flint's formulation could not be used for gases of equal W values, an alternate (also algebraic) formulation has been empirically derived and is used in Table 6 [see UNISCIENCE - H2 gas calculations], i.e., interdiffusion is approximated as the square-root of the quotient of (a) the sum of calculated mobilities (each calculated as inverse-square-root of respective W) and (b) the product of their W values (Eq.6a), which may be expressed solely in terms of "W" (Eq.6b). Eq.6a:I = k x sq.root of [(M1+M2)/W1W2], when M = inverse-square-root of W Eq.6b:I = k[sq.rt.(W1sq.rt.W2)+(W2sq.rt.W1)]/W1W2 ;k = 8.9585 adjusts all calculations to a base of .181 for N2-O2. Table 6 lists ten gas-pairs, i.e. all combinations of the 5 gases H2, N2, O2, CO, and CO2 (Col.1) and their observed coefficients of interdiffusion (Col.2).

Results of calculations using Equation 6 (i.e. all values are multiples of Z) are shown in Col.3. For gas-pairs which include H2, calculations in Col. 4 are identically derived except that H2 is here arbitrarily assigned a value of 2 relative to 32 for O2, 28 for N2 and CO and 44 for CO2, all approximating conventionally-appraised molecular weight values. Also shown in Table 6, Col.5, are Hirshfelder, Curtiss and Bird's approximations derived from viscosity data.*

Relative to observed values shown in Col. 2, approximations based on atomic number (Col.3) are preferable to those based on conventionally-appraised atomic weight values (Col.4); and comparable thus preferable to less-directly derived calculations based on viscosity (Col.5) in accord with Newton's assertion that "More is in vain when less will serve".

This result further illustrates the utility of Flint's methodology, and also may help explain various anomalies concerning hydrogen as noted by:

(1) Arrhenius, who referred to hydrogen's cathode ray absorption-to-density ratio (5610) as a "notable exception" from the mean (2791) for 8 solids (collodium, paper, glass, mica, aluminum, brass, silver, gold) and 2 gases at 760 mm. Hg. (air, SO2) as derived by Lenard;* whereas when the density of H2 is based on its "W" value as per assumed to be about twice as dense as conventionally appraised, its absorption-to-density ratio (approx. 2800) correlates closely with the mean for the other ten substances.

(2) Graham, who commented that the "want of mechanical equivalency in hydrogen mixtures is exceedingly remarkable, being a marked departure from the usual uniformity of gaseous properties"; whereas as shown in Table 6b, calculations assuming interdiffusion into air, v.s. diffusion into a vacuum, and Flint's W values v.s. conventionally-appraised values yield preferable results.

Table 6b, Time of Passage;Diffusion vs Interdiffusion

	Observed	Diffusion	Interdiff.
O2	1.0000	1.0000	1.0000
H2	.2631	.2502	.2571
O2/H2	.7255	.6251	.6955

(3) dissociative ionization of water vapor, branching ratio for hydrogen; whereas ...

(4) the observation that an average of less than 4% separates diffusion coefficients for He and H2 in the 8 instances affording direct comparision listed in cited references

(5) and last, but certainly not least, the fact that H2 gas and He gas have essentially the same lifting power is a most direct and definitive way of nature saying that, at least in gaseous form in the fashion we may use them to fill balloons, H2 AND HE WEIGH THE SAME.

[N]

Observations on Behavior of H2 Gas. S.H. Shakman

Flint's hypothesis of a weight of 4 for H2 gas on a scale of 32 for O2 gas* is supported by
 (1) similarities with He gas:
   (a) He has about 98% the lifting power of H2** and
   (b) an average of less than 4% separates observed diffusion coefficients for He and H2 in the 8 instances affording direct comparison listed in references***;
 (2) peculiarities in the behavior of H2 gas as noted by
   (a) Arrhenius, who referred to hydrogen's cathode ray absorption-to-density ratio (5610) as a "notable exception" from the mean [2794=mean for 8 sol- ids (collodium, paper, glass, mica, aluminum, brass, silver, gold) and 3 gases at 760 mm. Hg. (H2, air, SO2) as derived by Lenard]; however, when H2 is assigned a weight of 4 (v.s. 2), its absorption-to-density ratio (2805) correlates closely with this mean****, and
   (b) Graham, who commented that the "want of mechanical equivalency in hydrogen mixtures is exceedingly remarkable, being a marked departure from the usual uniformity of gaseous properties"*****; and
 (3) previously reported studies involving approximately 200 diffusion coefficients (see [O] The Weight of Hydrogen Gas.)******

*FLINT, L.H., Behavior Patterns of Hydration (1964), Ch. 11.
**CRC Handbook of Chemistry and Physics, 1985-6, B-20.
***CHAPMAN, S., etal, Mathematical Theory of Non-Uniform
    Gases (1970), 263, 267; and HIRSCHFELDER, J.O., etal.,
    Molecular Theory of Gases and Liquids (1954), 579, 601.
****ARRHENIUS, S., Theories of Chemistry (1907), 91.
*****GRAHAM, THOMAS, Elements of Chemistry (1850), 81.
******SHAKMAN, S.H., Abstracts AAAS, 1986, p. 119 (No. 212).

PRESS RELEASE -- NUMBERS ANALYST OFFERS "IRREFUTABLE" PROOF OF REVOLUTIONARY SCIENTIFIC HYPOTHESIS

In a paper presented to the 1986 AAAS Convention (on May 27, 1986), numbers analyst Stuart Hale Shakman has presented what he claims is irrefutable evidence in support of Lewis H. Flint's controversial hypothesis concerning atomic weights. This evidence shows that hydrogen gas weighs four on a scale of 32 for oxygen gas, thus addressing a major obstacle to further consideration of Flint's system.

Says Shakman, "Helium gas weighs four. Everyone agrees on that. And according to the 1985-6 CRC HANDBOOK, helium 'has about 98% the lifting power of hydrogen'. This in itself is pretty strong evidence that hydrogen gas also weighs four.

To test this hypothesis, Shakman compiled an exhaustive listing of avaiable interdiffusion measurements (the rates at which gases mix). Building on Lewis Flint's 1964 attempt to extend Sir Thomas Graham's (18th Century) law of diffusion (one gas) into the realm of interdiffusion, Shakman devised a formula whereby he could approximate interdiffusion behavior using only atomic numbers.

He then divided the interdiffusion measurements into groups, each with two or more gases interdiffusing with a single common gas, and compared the results for hydrogen gas with those for other gases in each group.

In 100 percent of 48 groups of gases, the result supported a value of four for H2 gas in preference to a avalue of two. Says Shakman of this result: "100 percent is a strong vote of support for Flint's hypothesis. This also by the way constitutes support for Prout's hypothesis and further, seems to clearly establish the hydrogen atom itself as the pure and simple basic building block of all matter."

As Shakman points out in his paper, Albert Einstein had been convinced of the existence of an algebraic system underlying reality. "With all due respect to the legions of scientists who have chosen to ignore the studied conclusion of Einstein, including the "high energy physics" purveyors of a sledgehammer approach to discovery, the facts clearly support Einstein's contention and Flint's hypothesis. In other words, EINSTEIN WAS RIGHT!"

[O]

The Weight of Hydrogen Gas (H2). S.H. Shakman

A study involving approximately 200 interdiffusing gas-pairs supports a value of 4 for the weight of H2 gas in preference to a value of 2, on a scale of 32 for O2 gas. Table A lists ten illustrative gas-pairs (A-B); observed diffusion coefficients (Do)*; calculations (Dc) based on viscosity data and conventionally-appraised molecular weight values*; calculations (Dc') based on atomic numbers per equation Dc'= 8.9585sqrt(WAsqrtWB+WBsqrtWA)/WAWB, when gases (A,B) are assigned weight values (WA,WB) equal to double the sums of atomic numbers of component atoms, and 8.9585 adjusts raw calculations relative to the base of .181 for N2-O2; and alternate calculations (Dc") identically derived as are Dc' except that H2 is arbitrarily assigned a WA value of 2 (thus WA and WB values approximate conventionally-appraised weight values). Note that orders of agreement with observed values exhibited by Dc' values (based on a value of 4 for the weight of H2 relative to 32 for (2) are (a) comparable to those exhibited by the less-directly derived Dc values and (b) preferable to those exhibited by Dc" values.

TABLE A-OBSERVED vs CALCULATED DIFFUSION COEFFICIENTS

A-B	Do	Dc	Dc'	Dc"
H2-N2	.674	.656	.703	1.133
N2-O2	.181	.175	.181	-BASE
N2-CO2	.144	.130	.149	
H2-CO	.651	.661	.703	1.133
CO-N2	.192	.174	.197	
CO-O2	.185	.175	.181	
H2-O2	.697	.689	.651	1.053
O2-CO2	.139	.128	.137	
H2-CO2	.550	.544	.545	.885
CO2-CO	.137	.128	.149	

*HIRSCHFELDER, J.O., CURTISS, C.F., and BIRD, R.B., Molecular Theory of Gases and Liquids (1954), p. 579.

(As Newton affirmed,"More is in vain when less will serve.")

Relative D_o/D_c; values = ~ 0.5 are for "weight" value of 2 for H2, from below table, in contrast to values =~ 1.0 for "weight value of 4 for H2.

See below tables for comparative calculations for 48 sets of gas pairs, comparing H2=2 versus H2=4 with calculations with other gases versus given base gases.

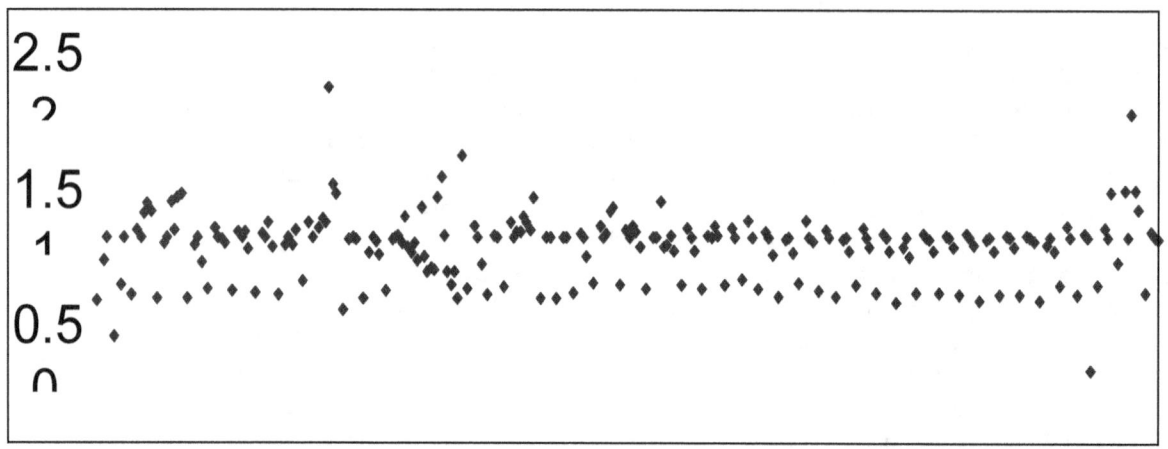

Interdiffusion calculations comparing conventional "weight" values for H2 gas (= 2) to
Z X 2 (= 4) for 48 groups of gas pairs

Dc=8.9585*(SQRT((col.D*SQRT(col.E)+col.E*SQRT(col.D))))/(col.D*col.E); relative Do/Dc=(Do//Dc)/base
Do/Dc for each group-- usually second value listed

For items designated *H2, W=Z=2 in accord with conventional "weight" of one for the Hydrogen
 atom.

			W(a)=2Z	W(b)=2Z	Do	Dc	Do/Dc	relative Do/Dc	Ref-value/temp
1	*H2	*H2	2	2	1.285	5.32675597	0.241235004	0.415596882	
	H2	H2	4	4	1.285	2.239625	0.573756767	0.988461538	5b-1.28,3f-
									1.285, 4b-1.29
	H2	He	4	4	1.3	2.239625	0.580454317	1	3a
2	*H2	*H2	2	2	1.455	5.32675597	0.273149363	0.393409738	
	H2	H2	4	4	1.455	2.239625	0.649662332	0.935691318	4c/23 deg C
	He	He	4	4	1.555	2.239625	0.694312664	1	"
3	*He	*H2	4	2	1.3	3.47987104	0.373577063	0.538053073	
	He	H2	4	4	1.3	2.239625	0.580454317	0.836012862	4a
	He	He	4	4	1.555	2.23962	0.694312664	1	4c/23 deg C
4	*D2	*H2	4	2	1.16	3.47987104	0.333345687	0.275691474	
	D2	H2	8	4	1.16	1.46310554	0.792834125	0.655708526	4a-1.12, 2a-1.2
	D2	CO2	8	44	0.41	0.33908795	1.209125844	1	4a
5	*CH4	*H2	16	2	0.625	1.54931778	0.403403361	0.586590591	
	CH4	H2	20	4	0.625	0.85200398	0.733564648	1.066679564	1a
	CH4	CH4	20	20	0.206	0.29954550	0.687708543	1	4c
	CH4	air	20	28.8	0.196	0.23899415	0.820103748	1.192516447	2a
	CH4	O2	20	32	0.194	0.22406983	0.865801506	1.258965757	4a
	CH4	CO2	20	44	0.153	0.18477368	0.82804	1.204056586	1a
6	*NH3	*H2	17	2	0.645	1.49718645	0.430808065	0.551955167	
	NH3	H2	20	4	0.645	0.85200398	0.757038717	0.969924812	4a
	NH3	He	20	4	0.665	0.85200398	0.780512786	1	4a
	NH3	Ne	20	20	0.296	0.29954550	0.988163731	1.266044258	4a
	NH3	A	20	36	0.172	0.20857083	0.824659916	1.0565617	4a
	NH3	Kr	20	72	0.14	0.13795030	1.014858243	1.300245507	4a
	NH3	Xe	20	108	0.113	0.10901074	1.036595146	1.328095023	4a
7	*H2O	*H2	18	2	0.749	1.44975935	0.51663747	0.561242428	
	H2O	H2	20	4	0.749	0.85200398	0.879103874	0.955003115	1b-.7516,2a-.747
	H2O	air	20	28.8	0.22	0.23899415	0.920524615	1	1b-.7516,2a-.747
	H2O	CO2	20	44	0.139	0.18477368	0.752271634	0.817220552	3c-.1384,1b-.1387
8	*H20	*H2	18	2	0.965	1.44975935	0.665627715	0.630130514	
	H2O	H2	20	4	0.965	0.85200398	1.132623817	1.072222222	7b-.91,.102/34.4C
	H2O	He	20	4	0.9	0.85200398	1.056333094	1	7b/34.0 C
	H2O	N2	20	28	0.256	0.24316447	1.052785368	0.996641471	7b/34.4 C
	H2O	CO2	20	44	0.188	0.18477368	1.017460915	0.96320083	7b-.174,.202/34.4C
9	*H2O	*H2	18	2	1.055	1.44975935	0.727706983	0.613870541	
	H2O	H2	20	4	1.055	0.85200398	1.238257126	1.044554455	7b1.12,.99/55.5C
	H2O	He	20	4	1.01	0.85200398	1.185440472	1	7b/55.3 C
	H2O	N2	20	28	0.303	0.24316447	1.246070181	1.051145301	7b/55.4 C
	H2O	CO2	20	44	0.202	0.18477368	1.093229281	0.922213563	7b192,.211/55.5C

#	Gas1	Gas2	W	Z					Notes
10	*H2O	*H2	18	2	1.15	1.449759346	0.7932351	0.603428092	
	H2O	H2	20	4	1.15	0.85200398	1.349758953	1.026785714	7b-1.1,1.2/79.2 C
	H2O	He	20	4	1.12	0.85200398	1.31454785	1	7b/79.3 C
	H2O	N2	20	28	0.359	0.243164474	1.476366981	1.1230987	7b/79.0 C
	H2O	CO2	20	44	0.228	0.184773682	1.233941961	0.93868166	7b-.215,.241/79.2C
11	*CO	*H2	28	2	0.651	1.133226866	0.574465731	0.588526001	
	CO	H2	28	4	0.651	0.702635841	0.926511234	0.949187953	1a
	CO	N2	28	28	0.192	0.196699274	0.976109348	1	2a
	CO	CO	28	28	0.1825	0.196699274	0.927812271	0.950520833	4b-.175,4c-.190
	CO	O2	28	32	0.1865	0.181000079	1.030386293	1.055605394	1a-.185,4a-.188
12	*N2	*H2	28	2	0.676	1.133226866	0.596526627	0.690512337	
	N2	H2	28	4	0.676	0.702635841	0.962091542	1.113673806	1a-.674,4a-.678
	N2	He	28	4	0.607	0.702635841	0.863889891	1	4a
	Na-N2 (W=2Z=28 & 22)are excluded due to high calc.(=103.16); calcs/discussion at end of table								
	N2	N2	28	28	0.182	0.196699274	0.92527032	1.071051218	4c-.178,6-.185
	N2	A	28	36	0.167	0.168273516	0.99243187	1.148794402	4a
	N2	CO2	28	44	0.144	0.148765405	0.967966982	1.12047495	2a
	N2	Cd	28	96	0.17	0.093218095	1.82368027	2.111010083	3d/15 C
	N2	Xe	28	108	0.105	0.087001093	1.206881388	1.397031498	4a
	N2	I2	28	212	0.068	0.059022007	1.152112629	1.33363365	1b-.0654,2a-.07
13	*N2	*H2	18	2	0.743	1.449759346	0.512498852	0.477763754	
	N2	H2	28	4	0.743	0.702635841	1.057446769	0.985777306	3a/15 C
	N2	CO	28	28	0.211	0.196699274	1.072703502	1	"
	N2	CO2	28	44	0.158	0.148765405	1.062074883	0.990091745	"
14	*N2	*H2	28	2	0.76	1.133226866	0.670651238	0.551763303	
	N2	H2	28	4	0.76	0.702635841	1.081641379	0.889896248	5a/20.2 C
	N2	O2	28	32	0.22	0.181000079	1.215469086	1	"
	N2	A	28	36	0.2	0.168273516	1.188541162	0.977845653	"
	N2	CO2	28	44	0.16	0.148765405	1.075518869	0.88485909	"
15	*air	*H2	28.8	2	0.611	1.11572755	0.547624731	0.611784623	
	air	H2	28.8	4	0.611	0.691478047	0.883614458	0.987139015	1a
	air	HCN	28.8	28	0.173	0.193268736	0.895126669	1	3b
	air	CNCl	28.8	60	0.111	0.121037721	0.917069478	1.024513636	"
	air	C4H11N	28.8	84	0.085	0.099001072	0.858576563	0.959167672	1c-.821 -.0884
	air	COCl2	28.8	96	0.095	0.09149595	1.038297321	1.159944572	3b
	air	C6H5NH2	28.8	100	0.075	0.089326292	0.839618422	0.937988388	3b/30C, 1c
	air	C7H8	28.8	100	0.071	0.089326292	0.794838773	0.887962341	1c.0821 -.0884
	air	C6H5Cl	28.8	104	0.075	0.08729411	0.859164499	0.95982449	3b/30 C
	air	C8H10	28.8	116	0.061	0.081895002	0.744856202	0.832123796	1c-.056 -.0658
	air	C3H7Br	28.8	120	0.088	0.080293164	1.095983717	1.224389525	1c-.085 -.0902
	air	C7H7Cl	28.8	132	0.059	0.075968531	0.776637363	0.867628449	1c-.051-.066
	air	C8H18	28.8	132	0.051	0.075968531	0.671330602	0.749983913	1c
	air	C9H12	28.8	132	0.053	0.075968531	0.697657293	0.779395047	1c-.0481-.056
	air	C10H8	28.8	136	0.051	0.074666282	0.683039233	0.763064331	1c
	air	C3H7I	28.8	156	0.08	0.068983682	1.159694548	1.295564737	1c-.079-.0802
	air	CCl3NO2	28.8	160	0.088	0.06798705	1.294364139	1.446012262	3b/25 C
	air	C12H10	28.8	164	0.061	0.067029983	0.910040513	1.016661155	1c
	air	C10H10O2	28.8	172	0.044	0.065224762	0.674590427	0.75362566	1c-.0434-.0455

	air	C10H1202	28.8	176	0.038	0.064372229	0.59031667	0.659478362	1c
	air	C14H10	28.8	188	0.042	0.061992266	0.677503863	0.756880435	1c
	air	C12H12N2	28.8	196	0.03	0.060537594	0.495559833	0.553619784	1c
	air	I2	28.8	212	0.083	0.057897209	1.433575139	1.601533268	2a
16	*C2H4	*H2	28	2	0.556	1.133226866	0.490634326	0.625386279	
	C2H4	H2	32	4	0.556	0.651407885	0.853535877	1.087958175	1a-.486 ,41-.505,2a-.625
	C2H4	N2	32	28	0.142	0.181000079	0.784530046	1	4a
	C2H4	CO	32	28	0.116	0.181000079	0.6408837	0.816901408	1a
17	*C2H4	*H2	28	2	0.602	1.133226866	0.531226375	0.589889666	
	C2H4	H2	32	4	0.602	0.651407885	0.924152155	1.02620621	7a/25.2C
	C2H4	N2	32	28	0.163	0.181000079	0.900552095	1	
18	*O2	*H2	28	2	0.697	1.133226866	0.61505778	0.640021545	
	O2	H2	32	4	0.697	0.651407885	1.069990119	1.11341853	1a
	O2	He	32	4	0.626	0.651407885	0.96099543	1	4a
	O2	N2	32	28	0.181	0.181000079	0.999999566	1.040587223	1a
	O2	air	32	28.8	0.178	0.177822444	1.000998503	1.041626705	1a
	O2	O2	32	32	0.185	0.166461124	1.111370604	1.156478553	4c-.181,4b-.189
	O2	A	32	36	0.166	0.154680876	1.073177271	1.116735041	4a
	O2	CO2	32	44	0.139	0.136634724	1.017310946	1.058601232	4a-.138,1a-.139
	O2	Xe	32	108	0.099	0.079621644	1.243380511	1.293846435	4a
19	*C2H6	*H2	30	2	0.459	1.090879606	0.420761372	0.561928535	
	C2H6	H2	36	4	0.459	0.609548718	0.753016102	1.005656087	1a-.459,4a-.46
	C2H6	N2	36	28	0.126	0.168273516	0.748780932	1	4a
20	*C2H6	*H2	30	2	0.537	1.090879606	0.492263305	0.559695115	
	C2H6	H2	36	4	0.537	0.609548718	0.880979623	1.001659044	7a/25.2C
	C2H6	N2	36	28	0.148	0.168273516	0.87952046	1	"
21	*CH4O	*H2	32	2	0.506	1.052792588	0.480626484	0.597358019	
	CH4O	H2	36	4	0.506	0.609548718	0.830122326	1.031737212	1c
	CH4O	air	36	28.8	0.133	0.165302203	0.804586979	1	"
	CH4O	CO2	36	44	0.088	0.126816702	0.69391491	0.862448595	"
22	*A	*H2	36	2	0.7	0.986889371	0.709299361	0.674496905	
	A	H2	36	4	0.7	0.609548718	1.14839057	1.092043682	4a
	A	He	36	4	0.641	0.609548718	1.05159765	1	1a
	A	A	36	36	0.157	0.143672011	1.092766776	1.039149123	6-.156,4c-.157,2b-.158
	A	Kr	36	72	0.119	0.093858264	1.267869176	1.205659955	4a
	A	Xe	36	108	0.095	0.073664902	1.289623649	1.226347024	4a
23	*A	*H2	36	2	0.77	0.986889371	0.780229297	0.656459635	
	A	H2	36	4	0.77	0.609548718	1.263229627	1.062840453	2a/20 C
	A	N2	36	28	0.2	0.168273516	1.188541162	1	"
	A	O2	36	32	0.2	0.154680876	1.292984664	1.087875377	"
	A	A	36	36	0.18	0.143672011	1.252853629	1.054110425	2b/22.3 C
	A	CO2	36	44	0.14	0.126816702	1.103955538	0.928832399	2a/20 C

									.0692,1b-.097
24	*CO2	*H2	44	2	0.546	0.884510091	0.617290866	0.622707413	
	CO2	H2	44	4	0.546	0.544738343	1.002316079	1.011111111	1a-.535,5a-.550,4a-.557
	CO2	He	44	4	0.54	0.544738343	0.991301617	1	4a
	CO2	Ne	44	20	0.232	0.184773682	1.255590066	1.266607503	4a
	CO2	CO	44	28	0.137	0.148765405	0.920913031	0.928993775	1a
	CO2	air	44	28.8	0.138	0.146112933	0.944474914	0.952762407	1a
	CO2	A	44	36	0.128	0.126816702	1.009330778	1.018187362	4a
	CO2	CO2	44	44	0.10025	0.111798107	0.896705701	0.904574033	4c-.0965,4b-.104
25	*CO2	*H2	44	2	0.619	0.884510091	0.699822428	0.658920044	
	CO2	H2	44	4	0.619	0.544738343	1.136325372	1.069910785	3a/15 C
	CO2	N2	44	28	0.158	0.148765405	1.062074883	1	"
	CO2	N2O	44	44	0.107	0.111798107	0.957082394	0.901143987	"
26	*CO2	*H2	44	2	0.6	0.884510091	0.678341611	0.630711028	
	CO2	H2	44	4	0.6	0.544738343	1.101446241	1.02410685	5a/20.2C
	CO2	N2	44	28	0.16	0.148765405	1.075518869	1	"
	CO2	O2	44	32	0.16	0.136634724	1.171005405	1.088781833	"
	CO2	A	44	36	0.14	0.126816702	1.103955538	1.026439954	"
27	*CO2	*H2	44	2	0.646	0.884510091	0.730347801	0.658487796	
	CO2	H2	44	4	0.646	0.544738343	1.185890453	1.06920893	7a/25.2C
	CO2	N2	44	28	0.165	0.148765405	1.109128833	1	"
28	*N2O	*H2	44	2	0.535	0.884510091	0.604854603	0.697129893	
	N2O	H2	44	4	0.535	0.544738343	0.982122898	1.131953412	1a
	N2O	CO2	44	44	0.097	0.111798107	0.867635442	1	1a-.096,4a-.098
29	*CH2O2	*H2	46	2	0.5104	0.863395544	0.591154314	0.626200289	
	CH2O2	H2	48	4	0.5104	0.518968992	0.983488431	1.04179353	1c
	CH2O2	air	48	28.8	0.1308	0.138554343	0.944033921	1	1c
	CH2O2	CO2	48	44	0.0874	0.105893175	0.825360083	0.874290707	1c
30	*C2H6O	*H2	46	2	0.37615	0.863395544	0.435663588	0.572102242	
	C2H6O	H2	52	4	0.37615	0.496413525	0.757735197	0.995038413	1c-.3753,3c-.377
	C2H6O	air	52	28.8	0.1005	0.131974021	0.761513512	1	3c-.099,1c-.102
	C2H6O	CO2	52	44	0.06855	0.100758015	0.680342901	0.893408837	1c-.0685,3c-.0686
31	*C2H4O2	*H2	60	2	0.4163	0.7478274	0.556679255	0.669569263	
	C2H4O2	H2	64	4	0.4163	0.442644757	0.940483297	1.131205631	1c
	C2H4O2	air	64	28.8	0.0968	0.11643025	0.831399059	1	1c-.0872-.1064
	C2H4O2	CO2	64	44	0.0716	0.088650502	0.80766604	0.971454119	1c
32	*C4H8	*H2	56	2	0.378	0.776169318	0.487007141	0.607897781	
	C4H8	H2	64	4	0.378	0.442644757	0.85395793	1.065937412	7a/15.2 C
	C4H8	N2	64	28	0.095	0.118582017	0.801133275	1	"
33	*C3H8O	*H2	60	2	0.3153	0.7478274	0.421621353	0.567595728	
	C3H8O	H2	68	4	0.3153	0.428144405	0.736433774	0.991402977	1c
	C3H8O	air	68	28.8	0.0834	0.112274873	0.742819814	1	1c-.0818-.085
	C3H8O	CO2	68	44	0.0577	0.085419633	0.675488733	0.909357452	1c

34	*CS2	*H2	76	2	0.3689	0.658719383	0.560026028	0.659618261
	CS2	H2	76	4	0.3689	0.402840282	0.915747546	1.078599517 1c
	CS2	air	76	28.8	0.0892	0.105062883	0.849015349	1 1c
	CS2	CO2	76	44	0.063	0.079818691	0.78928881	0.929651991 1c
35	*C3H6O2	*H2	74	2	0.33135	0.668186737	0.495894309	0.605576081
	C3H6O2	H2	80	4	0.33135	0.391711437	0.845903308	1.032999978 1c-.3297-.3330
	C3H6O2	air	80	28.8	0.08345	0.101907449	0.818880277	1 1c-.0829-.0840
	C3H6O2	CO2	80	44	0.05775	0.07737087	0.746404946	0.911494595 1c-.0567-.0588
36	*C4H10O	*H2	74	2	0.2853	0.668186737	0.426976449	0.537458691
	C4H10O	H2	84	4	0.2853	0.381428507	0.747977654	0.941520528 1c-.2716-.2964,3c-.299
	C4H10O	air	84	28.8	0.07865	0.099001072	0.794435843	1 1c-.0703,3c-.0786
	C4H10O	CO2	84	44	0.05142	0.075117806	0.684591351	0.861732709 1c-.0476-.05525,3c-.0541
37	*C6H6	*H2	78	2	0.3064	0.649633606	0.471650477	0.606414321
	C6H6	H2	84	4	0.3064	0.381428507	0.803296016	1.032820345 1c-.2948,3c-.318
	C6H6	air	84	28.8	0.077	0.099001072	0.777769357	1 1c-.2948,3c-.318
	C6H6	O2	84	32	0.0715	0.09238827	0.773907766	0.995035044 1c-.0633,3c-.0797
	C6H6	CO2	84	44	0.0528	0.075117806	0.702895933	0.903733127 1c
38	*C4H8O2	*H2	88	2	0.27945	0.609134261	0.458765855	0.597511997
	C4H8O2	H2	96	4	0.27945	0.354723794	0.78779604	1.026051918 1c-.264-.2949
	C4H8O2	air	96	28.8	0.07025	0.09149595	0.767793545	1 1c-.067-.0735
	C4H8O2	CO2	96	44	0.04995	0.069307003	0.72070639	0.938672114 1c-.0471-.0528
39	*C5H12O	*H2	88	2	0.23445	0.609134261	0.384890516	0.585704301
	C5H12O	H2	100	4	0.23445	0.346961213	0.675723946	1.028277926 1c-.234-.2349
	C5H12O	air	100	28.8	0.0587	0.089326292	0.657141351	1 1c-.0585-.0589
	C5H12O	CO2	100	44	0.04205	0.067629214	0.621772711	0.946178033 1c-.0419-.0422
40	*C5H10O2	*H2	92	2	0.2346	0.594886688	0.394360817	0.547127556
	C5H10O2	H2	112	4	0.2346	0.326338998	0.718884354	0.997364401 1c-.2123-.2569
	C5H10O2	air	112	28.8	0.06025	0.08358953	0.720784051	1 1c-.050-.0705
	C5H10O2	CO2	112	44	0.04135	0.06319775	0.654295448	0.907755169 1c-.0376-.0451
41	*C6H14O	*H2	102	2	0.1997	0.563145316	0.354615397	0.581988549
	C6H14O	H2	116	4	0.1997	0.320219567	0.623634595	1.023498121 1c
	C6H14O	air	116	28.8	0.0499	0.081895002	0.609316795	1 1c
	C6H14O	CO2	116	44	0.0351	0.061890157	0.5671338	0.930770011 1c
42	*C6H12O2	*H2	116	2	0.22395	0.526046257	0.425723018	0.592157664
	C6H12O2	H2	128	4	0.22395	0.303691299	0.737426461	1.025720276 1c-.2115-.2364
	C6H12O2	air	128	28.8	0.0556	0.077336589	0.718935249	1 1c-.050-.0612
	C6H12O2	CO2	128	44	0.041	0.05837592	0.702344392	0.976923017 1c-.0395-.0425
43	*C7H14O2	*H2	130	2	0.20745	0.495314442	0.418824856	0.549106838
	C7H14O2	H2	144	4	0.20745	0.285090308	0.727664162	0.95401541 1c-.2029-.212
	C7H14O2	air	144	28.8	0.0551	0.072239712	0.762738373	1 1c-.0512-.059
	C7H14O2	CO2	144	44	0.0376	0.05445255	0.690509444	0.905303141 1c-.0364-.0388
44	*CCl4	*H2	148	2	0.293	0.462607021	0.633366954	0.659716967
	CCl4	H2	148	4	0.293	0.280937916	1.042935052	1.086324357 2a
	CCl4	O2	148	32	0.0636	0.066245956	0.96005861	1 2a

45	*C8H16O2	*H2	144	2	0.1882	0.46932484	0.401001575	0.589468418	
	C8H16O2	H2	160	4	0.1882	0.269467223	0.698415183	1.02666353	
	1c-.1850-.1914								
	C8H16O2	air	160	28.8	0.04625	0.06798705	0.680276607	1	
	1c-.0457-.0468								
	C8H16O2	CO2	160	44	0.03455	1.835949258	0.018818603	0.027663163	1c-.0327-.0364
46	*Hg	*H2	160	2	0.53	0.444034596	1.193600689	0.655382945	
	Hg	H2	160	4	0.53	0.269467223	1.966844033	1.07995584	3d
	Hg	N2	160	28	0.1262	0.069293968	1.82122635	1	1b-.1124,2a-.1190,3d-.14
	Cd	N2	160	28	0.17	0.069293968	2.453316002	1.347068146	3d
47	*Rn	*H2	172	2	0.476	0.42749957	1.113451412	0.822451757	
	Rn	H2	172	4	0.476	0.259266425	1.835949258	1.356125356	3e/15 C
	Rn	He	172	4	0.351	0.259266425	1.353819726	1	"
	Rn	Ne	172	20	0.217	0.083639708	2.59446147	1.916400995	"
	Rn	air	172	28.8	0.12	0.065224762	1.839792073	1.358963855	"
	Rn	A	172	36	0.092	0.056110978	1.639607838	1.211097612	"
48	*C9H18O2	*H2	158	2	0.172	0.446977831	0.384806557	0.601234367	
	C9H18O2	H2	176	4	0.172	0.256109408	0.671587979	1.049311053	
	1c-.171-.1730								
	C9H18O2	air	176	28.8	0.0412	0.064372229	0.640027547	1	1c-.040-.0424
	C9H18O2	CO2	176	44	0.0308	0.04841	0.636232178	0.994069991	1c
	N2	Na	28	22	20.4	0.228905786	89.11963441	103.160872	3d/15 C

The extraordinarily high calculated value for N2-Na may be taken to indicate an extremely small particle coming off the Na ion, in accord with Lenard.

References:
1. International Critical Tables, Vol. 5, National Research Council, McGraw Hill, 1929
a. p. 62, A Table, Diffusion of Gases into Gases
   b. p. 62, B Table, Diffusion of Vapors into Gases

   c. p. 63, C. Table.
2. American Institute of Physics Handbook, McGraw Hill, 1957.
   a. p.2-212, Table 2W-1, Diffusion Coefficients Do at Standard Temperature and Pressure.
   b. p. 2-213, Table 2W-4, Coefficients of Self-diffusion.
3. Jost, W., Diffusion in Solids, Liquids, Gases; Academic Pres Inc., 1960.
   a. p. 409, Diffusion Coefficients for 15 degrees C and 760 mm Hg.
   b. p. 412, Table IV.
   c. p. 412, Table V.
   d. p. 413, Table VI.
   e. p. 413, Table VII.
   f. p. 430, Table XV.
4. Chapman, Sydney and T.G. Cowling, The Mathematical Theory of Non-Uniform Gases; Cambridge University Press, 1970.
   a. p. 263, Table 23, Coefficients of diffusion at STP.
   b. p. 266, Table 24, Coefficients of self-diffusion.
   c. p. 267, Table 25, Self-diffusion coefficients from isotopic diffusion experiments
5. Hirschfelder, J.O., Bird, R.B., and Spotz, E.L., Chem Rev. 44, 205)1949).
   a. p. 209, Table 2.
   b. p. 226, Table 9.
6. Grew, K.E. and Gibbs, T.L., Thermal Diffusion in Gases, Cambridge University Press, 1952.
7. Hirschfelder, J.O., Curtiss,C.F.,,, and Bird, R.B., Molecular Theory of Gases and Liquids, John Wiley and Sons, Inc., N.Y., 1954.
   a. p. 579, Table 8.4-12, Comparison of Calculated and Observed Diffusion Coefficients.
   b. p. 601, Table 8.6-2, Coefficient of Diffusion for H2O and some Non-polar Gases.

[P]

     Reconciliation of contemporarily-appraised value for
weight of H2 gas (2) with Flint's alternate (4). S.H. Shakman
Flint presumed water units involved in hydration to be negatively-
charged ions (H2O-) weighing 18 each (v.s. 20 for a neutral H2O)*;
a like value of 18 figures prominently in conventional determinations
of weights of H2 and O2**.

As direct evidence that negative ions may weigh less than positive
ones, Flint cited greater mobilities of the former v.s. the latter,
and showed how these are calculable.*

Flint also projected "combining" wts. (average of neutral & anhydrous
wts.) of 1 for H-, 7 for Li+, 23 for Na+, & 39 for K+, which approximate
contemporary "weight" values, but he did not utilize these values in his
calculations.*

The hypothesis that an entity (e.g. water) may weigh other than the sum
of constituents has precedents in: Marignac's 1860 suggestion that a
grouping of primordial atoms in the form of a (larger) chemical atom
might weigh other than the sum of weights of constituent atoms***; and
in F. W. Aston's 1921 proposal of a "packing" effect which might allow
for an atomic nucleus to weigh less than the sum of wts. of constituent
"charges".***

*FLINT,L.H.,Behavior Patterns of Hydration,1964: 22,25,159+.
**CLARKE,F.W.,Memoirs N.A.S., 16(1922), p.11-31.
***PROUT,W. (1815-6),J.STAS & C.MARIGNAC(1860); Prout's Hypoth.:p.58,22.

Flint's use of values based on atomic numbers is consistent with contemporary recognition of the primacy of the atomic number[1] as established by Moseley[2]; contemporary science also embraces the concept of a shift in atomic-number-equivalent values in the case of an increase in the atomic number of some radioactive elements resulting from the loss of a nuclear electron[3].

1. Encyclopedia Britannica, 15th Ed. (1991), Vol.1, p. 676.
2. Moseley, H. G. J., Phil. Mag., 703 (1914).
3. Bertsch, G. F. and S. McGrayne, in Encyclopedia Britannica,
     15th Ed. (1991), Vol. 14, p. 330.

   As direct evidence that negative ions may weigh less than positive ones, Flint cited greater mobilities of the former v.s. the latter, and showed how these are calculable*,
i.e. as would be expected for such ions interdiffusing with air.

We may speculate that Flint was familiar with and may have borrowed from S.Arrhenius's (out-of-context) statements [THEORIES OF CHEMISTRY (1907),pp.100-1] that inherent in Mendeleev's scheme, "atomic weight ought to increase with the positive valence" [of consecutive elements] and that Thompson's scheme would require the supposition that a quantity of electricity "makes the difference between two consecutive elements".

Clear and dramatic evidence of change of weight with ionization may be found in the osmotic behavior of $(NH_4)_2MoO_4$ [Table 16, p. 68, FLINT (1964)]: whereas the neutral Mo with Z=42 would have H=4, Mo+6 with Z'=47 would have H=21 and then and only then would exhibit osmotic behavior as reported and calculated against by Flint. This at least would appear to be a most readily repeatable experiment.

And calculations for the SO_4^{--} ion in [E] "A Lock for Flint …" above provides direct evidence of a dramatic change in hydrational status. As for the H+ ion; conductivity clearly and precisely indicates an anhydrous state, whereas some volumetric and other data indicates that the H+ ion is fully hydrated.

ON THE WEIGHT OF THE ELECTRON
Georgi, Howard, <u>Science</u> **269**, 22 Sept. 1995, 1742, "Field Theory for Today", reviews Steven Weinberg"s <u>The Quantum Theory of Fields, Vol. I</u> notes that "We still have no idea why the electron should be so much lighter than all other charged particles. This is but one example of a deep mystery about the world at short distance."

[Q]

Primacy of Atomic #: Contemporary Applications. S.H. Shakman

1.DISSOCIATIVE IONIZATION OF WATER VAPOR: A modern model predicts a "branching ratio" for hydrogen of about half that observed(*1); which model coincidentally uses a contemporary weight value for hydrogen of about half that proposed by Flint(*2) and supported by recent studies(*3).

2.SOLUTE DIFFUSION (D) may be approximated(Dc) as inverse-sq. -root of Z' (or as inverse-square-root of relative anhydrous weight) (*2) in accord with Graham's law.

SOLUTE	D	Ref	Z	Dc	Error
Water*2	2.14	*4	9	2.14	BASE
NH3	1.96	*5	10	2.03	-.03
CO2	1.32	*5	22	1.37	-.04
Urea	1.18	*4	32	1.14	.04
SO2	1.23	*5	32	1.14	.08
H2SO4	1.06	*5	50	.91	.16
Sucrose	.52	*4	171	.48	.08

3.APPARENT EQUILIBRIUM DISTRIBUTION COEFFICIENTS (ED) of salts in silica gel water may be approximated (EDc) as sq.-root of Z' (or relative anhydrous wt.)(*2)

SALT	ED*6	Z	EDc	Error
HNO3	.493	32	.510	-.03
NaNO3	.519	42	.584	-.11
LiNO3	.534	34	.525	.02
NH4NO3	.577	42	.584	-.01
KNO3	.698	60	.698	BASE
RbNO3	.784	68	.743	.06
CsNO3	1.000	91	.860	.16

*1 J.H.Miller,etal.,J.CHEM.PHYS.(1 Ja 1987) p. 161, 157.

*2 FLINT,L.H.,Behavior Patterns of Hydration (1964), 21: Z'=Z + C (valence); Flint assumed water = H2O-, with Z'=9.

*3 SHAKMAN,S.H.,AAAS Abstracts 1986/#212, 1987/#110.

*4 ADAMSON, A.W., A Textbook of Physical Chem.(1979),p.370.

*5 LEAIST, D.G., J. Phys. Chem. (ACS), 1987, 91, 4635-38.

*6 DALTON etal.(1962),J.Colloid Sci.;per LING,G.N.(1984)274.

Proposed for 1988 AAAS Meeting, AAAS # 0925.16; withdrawn 25 Sept. 1987.

"Until the string people can interpret perceived properties of the real world, they simply are not doing physics. Should they be paid by universities and be permitted to pervert impressionable students?" Sheldon Glashow, The Sciences, May/June 1988, p. 25.

[R]

Hydration, Periodicity & Prime Numbers.   S.H. Shakman

(1) Groups of atomic numbers (Z) with near-equal hydrated Z values per Flint* (Zh) correlate with Mendeleev Groups(MG)** as in Fig.1 & 1a. Prime Z's in Fig.1,1a (bold) include 4 of total 8 prime numbers in MG1, 2 of 4 in MG2, and 2 of 3 in MG7.
(2) Symmetry of prime nos. w/in Flint's matrix is highlighted when no. 89 thru 1 are superimposed on series 1-89; 20 of 25 primes in each series would be superimposed on a prime in the other, at points circled in Fig.2a (89 on 1, 83 on 7, which points describe arcs 1-23-47-73,7-29-53-79... (Fig. 2b)
(3) First gap exceeding five between primes(89-97)
(a) encompasses 92, concluding atomic number of natural elements & Flint periods*;
(b) coincides w/the start of the "Actinides"** (89 is also Fibonacci no.); & (c) foreshadows larger gap: In extended Flint's matrix as in Fig.2c, light from above 37-41 through 89-97 would illuminate gap 113-127.

Fig. 1:

Z	Zh	MG
3	183	1
29	182	
55	181	
4	175	2
30	174	
56	173	
5	167	3
31	166	
57	165	
11	119	1
37	118	
17	71	7
43	70	

Figure 2:
Flint's matrix, !=prime.

```
 0.!!!.!.!...!.!...!.!...!23

23!.....!.!.....!...!.!...46

46.!.....!.....!.!.....!..69

69..!.!.....!...!.....!.......!97

92.....!...!.!.....!.!.....!.............!127
```

*   FLINT, L.H., Behavior Patterns of Hydration (1964), 21.

Figure 1b illustrates correlations between calculated hydrated "weights" (atomic number equivalents) and Mendeleev groups, suggesting the influence of gravitational attraction as underlying Mendeleev's periodicity.

Figure 1b. Hydrated Atomic-Number-Equiv. (Z'h)per Flint[3,4] v.s. Mendeleev Groups

1 Z'h (Z)	2 Z'h (Z)	3 Z'h (Z)	4 Z'h (Z)	5 Z'h (Z)	Mendeleev Group
		148(85)	212(77)	276(69)	
	85(**67**)	**149**(**59**)	213(51)	------------ (V)	
	86(**41**)	150(33)	214(25)		
23(**23**)	87(15)	**151**(**7**)			
	92(92)	156(84)	220(76)		

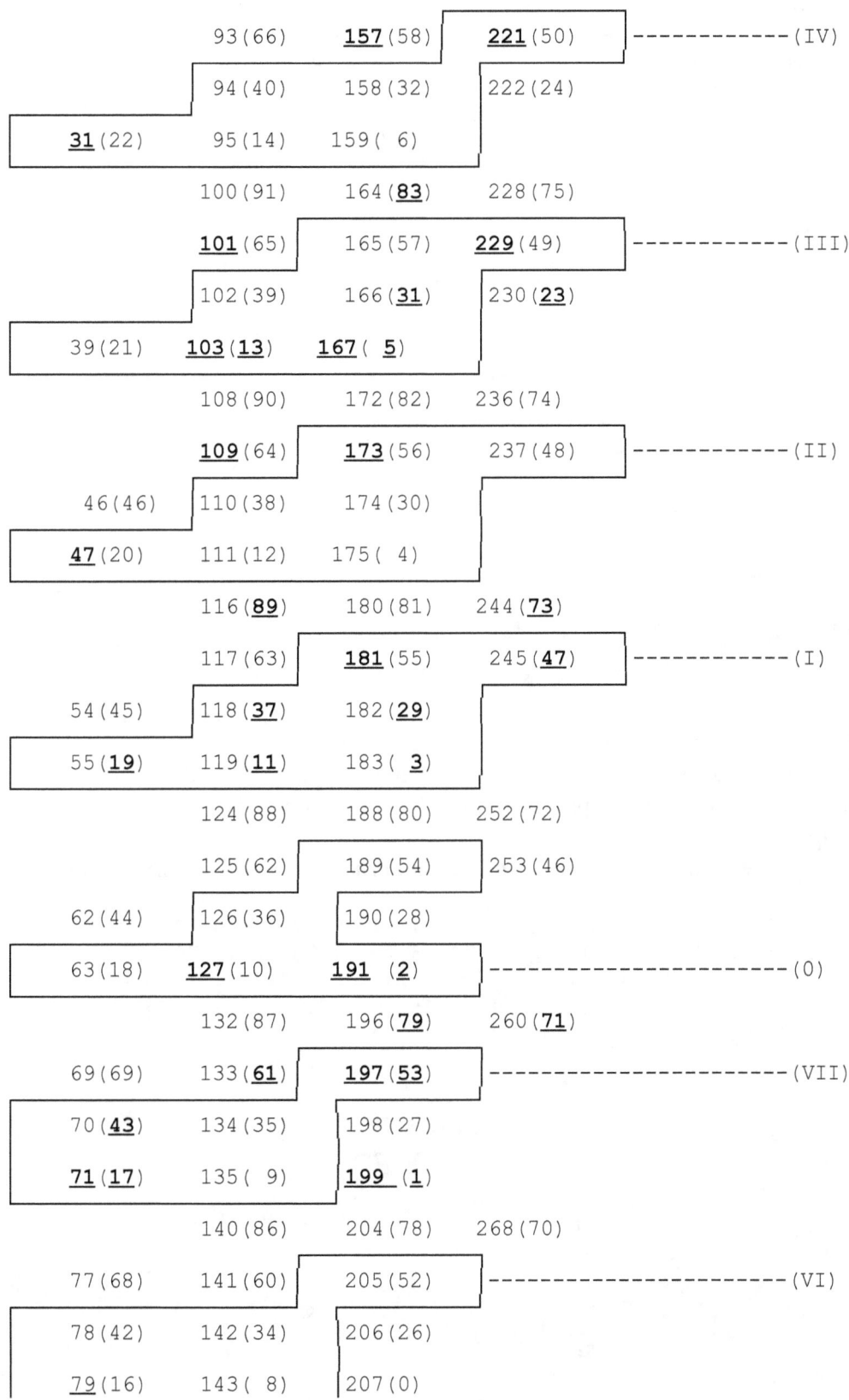

Underlined Bold = Prime numbers

```
Z    = Atomic number; Z'h  = Z + 9Hmax (Equation 3);
Hmax calculated as per eq. 2, assuming C=0 in eq. 1

Equation 1:  Z' = Z+C (Z=atomic number; C=valence)
Equation 2:  Hmax = 23n - Z', when Hmax = 23 to  0, and n = 1 to 4
                          (for Z' = 0 to 23, n=1;
                           for Z' = 23 to 46, n=2;
                           for Z' = 46 to 69, n=3;
                           for Z' = 69 to 92, n=4).
```

Figure 2a:

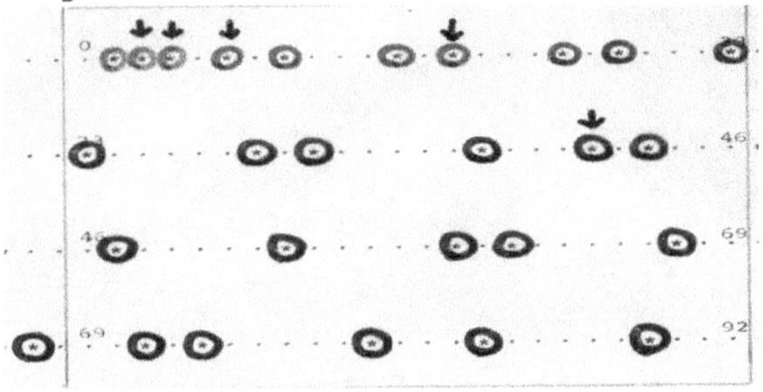

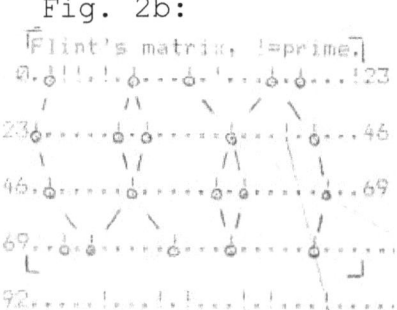

Note:

(1) In above Figure 2a, all prime numbers between zero and 92 are circled.

(2) To further illustrate the symmetry in this structure, as shown below, the first half is listed in sequence and the second backwards, with all prime numbers bold/underlined, beginning with numbers 1 and 89 juxtaposed.

(3) As shown, except for prime numbers 2, 3 and 5, and beyond prime number 5, all prime numbers line up in pairs:

```
     0  1  2  3  4  5  6  7  8  9 10 11 12 13 14 15 16 17 18 19 20 21 22
92 91 90 89 88 87 86 85 84 83 82 81 80 79 78 77 76 75 74 73 72 71 70 69 68

23 24 25 26 27 28 29 30 31 32 33 34 35 36 37 38 39 40 41 42 43 44 45
67 66 65 64 63 62 61 60 59 58 57 56 55 54 53 52 51 50 49 48 47 46
```

Quite a coincidence (at a minimum)!

Fig. 2b:

Figure 2c:

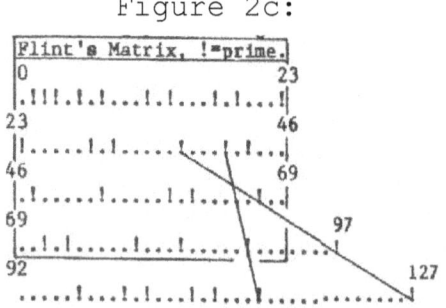

[S]

23 Skidoo: Hydrobiometric Correlation/Speculation. S.H. Shakman

1. 23=hydration no. at 0=atomic no. in
   (a) Flint's description [*1]; and
   (b) Berecz/Achs Balla graph [*2].
2. (a) 23 & 17 water molecules per 4 & 3 guests in Structure I &
   II gas hydrates, resp. [*3];
   (b) 17-23 glucose units in A.tumefaciens [*4]; &
   (c) 23=area, 17=circumf. of grt. circle of sphere w/area=92.
   [(17x17)/(23x4)=approx.pi)].
3. (a) 24 nos. per Flint period:0-23,23-46,46-69,69-92 [*1,*5];

   (b) "24-plet of fermions" in Russian lit.[*6];
   (c) 24 in Einstein's work on planetary motion [*7]; and
   (d) 24-25 4-dimensional equal spheres may touch a central one [*8].
4. (a) 6 is nearest whole-no. of water units per hydrate guest [2)(a)];
   (b) 6 equal circles around a central one/ hexagonal structure as in:
   (c) Flint's carotin [*9],
   (d) observed photosynthetic membrane [*10],
   (e) earthworm hemoglobin [*11], & (f) artificial cell [*12].

*1 FLINT,L.H., J.Washington Acad.Sci.22 (1932),99,212-3.
*2 BERECZ,E. ACHS-BALLA, Acta Chim.(Hung.)77(1973),277.
*3 VAN DER WAALS,J.N., Adv. in Chem. Phys.2(1959),11.
*4 MILLER, ETAL., Science(3 Jan.1986)7,48ff.
*5 SHAKMAN,S.H., Abstracts AAAS 1987 (#112).
*6 ZHELONKIN,A.V., ETAL., Soviet J.Nuc.Phys.34(1981),905-8.
*7 EINSTEIN,A., Meaning of Relativity, Princeton(1956)97.
*8 BUHLER, J.P., Science '85(Nov.)84-85.
*9 FLINT,L.H., Adv.Front.Plant Sci.6(1963)16.
*10 MILLER,K.R., Nature 300(4 Nov. 1982)53.
*11 CREWE,A.V., Science 221(22 July 1983)329.
*12 LE BON,G., Evolution of Matter(1905)244.

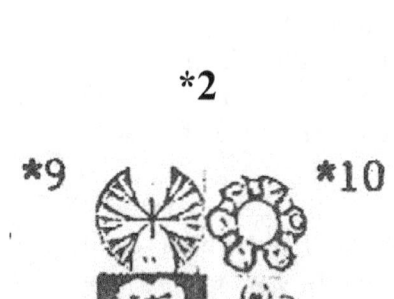

SPECULATIVE CORRELATIONS – CLATHRATES, PERIODICITY, PI :

Note that 17 is the number of cells in Structure II Hydrates and 23 is the number of cells in Structure I Hydrates

Surface area = 92; 92=4(pi)(r-square);r=sqrt of 7.32 = 2.70555; d=5.40; circumference = 17.0008 (See Shakman '86, figure 7.b.)

Note also that 17 + 6 = 23, and that 6 is a common feature in Figure 7.,e,h,i,h,k, and m.

When 92 represents the surface area of a sphere, 23 represents the area of a great circle and 17 is the circumference of that great circle. [When $23 = \pi r^2$, $17 = 2\pi r$; it may therefore be of interest to note that (17*17=289)/(23*4=92) = 3.1413, a close approximation to π.]

Wherefrom comes π? 3-1/7 to 3-10/71 = 22/7 to 223/71 = 3.1428571 to 3.140845; mid point = 3.141851

Sqrt(23/π) x 2π = 17.000779 Sqrt(23/X) x 2π = 17; X= (17)(17)/92 = 3.1413043 3.14159265/3.1413043 = 1.0000918; or X=/π - <.0001.

Notes on further (unsuccessful) search for algebraic derivation of π(pi):

[(23/3) = 7-2/3]/289 = .0265282 [289 + (23/3)/289]/92 = 3.1415926984604583215 = π (approx.) [3.1415926536 etc]: <[(17)(17)] + [23/3]/[(17)(17)]>/<(23)(4)> = π (approx.)

CONCLUSION OF STUDIES ON RELATIONSHIP BETWEEN PI, AND 17 AND 23:

While the numbers 17 and 23 may not be related to the derivation of pi, the close relationship <[(17)(17)]/[(23)(4)] = [π - .0003]; or π/((([(17)(17)]/92)) = 1.00009> may be a factor contributing to prominence of 17 and 23 in clathrate structures, as upper and lower limits on number of water molecules in mixed hydrates.

[When area of sphere surface equals 92, area of great circle = 23 = /r-square; therefore r = 2.7057581 and diameter (2/r) = 17.000779. 17.000779/17 = 1.0000458 or 1 + <(5/100,000)

ADDITIONAL NUMERICAL COINCIDENCES IN THE LITERATURE

SCIENCE, 3 Jan. 1986 p. 3, "Osmotic adaptation by bacteria", speaks of soil bacteria that contain complex sugars that are "osmoregulated, consist of 17 to 23 glucose units, and are localized within the periplasmic space."

Tie this to connection between clathrates and life, and symmetries of "23" and "17" as area and circumference of great circle of sphere with surface area of 92.

RESEARCHES ON THE ARSENIATES, PHOSPHATES, ETC. by Sir Thomas Graham, FRS (1833), p. 15, refers to "Theory of 23 atoms water", standing for from 23-24 atomic proportions of water corresponding with 50.82-49.75% water in subarseniate of soda, against "3 atoms of oxygen in the base" [3 : 23.457 in one experiment].

General result: Subarseniate of Soda

Experiment	Theory of 23 atoms water			
Arsenic acid	27.76	27.69		
Soda	22.85	22.55		
Water	50.22	49.75	100.83	100.

p. 45, refers to the concept of "affinity for water" and "anhydrous":

"...arsenic acid exhibits a weaker affinity for water than phosphoric acid, and is readily made anhydrous by heat."

F = M1M2/d-square; M1/M2 = (Fd-square/M2)/(Fd-square/M1)

[T]

Cosmological Speculation/Cycle of Gases. S.H. Shakman

As solutes in water exhibit behavior analogous to[*a] & possibly derivative of gases in a vacuum, so might the water medium itself relate to that of space; molecular mass of gas (or fluid density), inverse-sq.-root of which may be equated w/relative velocity (or velocity of efflux) [*b], to that value whose inverse-sq.-root may be equated to the velocity of light, which value may intimately relate to evolution of both electromagnetivity[*c] & matter[*d]; degassing of planets[*e] to cosmic outflow phenomena[*f]; inversion & ionization that may be associated with generation of biomass on earth[*g] (where volcanic gases impact rel. cold water), to that which may be associated with generation of protomass in the form of ionized hydrogen in star forming zones[*h] (where the hot outflow of an inverse black hole may impact our cell[*i] of cold space). An hypothesized attribute of zero mass for negatively ionized hydrogen[*j] may render it an attractive protomass candidate. On earth biomass formed of degassed methane etc. could be caught in sub-oceanic volcanic turbulence & cooked into petroleum[*k]. Degassed methane may have alternatively formed methane hydrates[*l] under permafrost & ocean floors in sufficient quantities to encircle earth w/a 40-mile thick CH_4 atmosphere[*m], which hydrates form a cap over still more methane underneath[*n].

*a-J.van't Hoff;b-T.Graham;c-L.Abbott;d-G.LeBon;e-R.Sparks, S.Brazier;f-W.Welch etal;C.Lada; g-J.Corliss;C.Ponnomperuma; h-R.Allen etal; i-H.Alfven; j-L.H.Flint; k-S.Fox; l-W.Rose; m-R.McIver; n-R.Malone. THANKS:H.Craig;I.Kaplan;S.Silverman.

EVOLUTION OF MATTER: Insofar as electromagnetism migrates at the speed of light and ionized hydrogen is reported as migrating at lower speeds, might protomass in the form of ionized hydrogen be considered electromagnetivity slowed down?

Such proposed protomass might be comprised of an influx of hot imponderable phenomena associated with regions of star formation (i.e. ionized hydrogen gas jets at 100-200km/sec., in accord with current understanding of materials in zones of star formation), imploding through the dimensional "barrier" of the cell that constitutes our ponderable universe, into the cold and compact medium of our beloved ponderable space, which protomass if initially comprised of negatively ionized hydrogen would possess zero weight thus zero mass, similar to that quality proposed by Mendeleev for element "X", 'Newtonium'; and which, in accord with Flint's work on combining weights, could also at its mid-point transition state comprise Mendeleev's element "Y", coronium or 'Flintium'.   REFERENCES:

ORIGIN OF PLANETS AND PLANETARY OUTGASSING, ETC.: Earth and other planets may be product of mass-energy outflows from a previously-young star (our sun), which planets may retain some "solar material" within the core and/or mantle which may both power and constitute some measure of degassed materials,

-which materials include water (which in any case is known to exist (at least 7 miles) deep in the earth, helium and methane,
-which methane, leaking out of various man-made holes, would logically contribute to increasing atmospheric methane,
-which degassing may comprise earthquakes and volcanoes and the source of natural gas, methane hydrate deposits and

ORIGIN OF LIFE: biomass (spontaneous hydrothermal biomass/hydrate generation, in accord with LeBon's suggestion of crystalline genesis), which last may immediately cook out into what is known as petroleum deposits.(involves the speculative correlation of gas hydrate structure with that of primitive organisms, with implications for study of origin of life [suggestive of spontaneous and continuous (hydrothermal) origin of life]. This work supports earlier work on this subject by [MM??

  More on analogy of biological and cosmic cells: Might ionization/hydration changes in solute ions incident to passage through semipermeable membrane of cell wall in biological arena be likened to ionization of primordial hydrogen incident to its entry to ponderable universe through cosmic cell wall?
  Flint, p. 40, "semipermeable" absorptive membranes were appraised as "permeable to anhydrous ions but not to hydrated ions. The appraisal involved the concept of water as of a greater molecular complexity than that represented by the symbol H_2O, but as yielding to the absorptive membrane matrix the negatively charged H_2O units involved in the hydration of solutes."

John Maddox on Origin of Life"\:

Nature 367, 409 (3 February 1994), "Origin of life by careful reading", reports Dr. Thomas Gold's "memorable stage-whisper, 'Bloody fool, they might be like rocks!', which was meant as a reference to the place of silicon as the next member after carbon in Group IV of the Periodic Table."
  ...
  ... Perhaps the next big effort, in the search for the origin of life, should be a careful reading of the journals in the hope of defining the latest dates at which such features [membrane channels] could have arisen."

[ʊ]

DIRECT CALCULATION OF HYDRATION NUMBERS FOR LIGHTER IONS
(*Nature SXA011**)

Hydration numbers calculated from conductance ----- three steps:
(a) total ionic weight is derived as inverse-square of conductance;
(b) weight of anhydrous solute is subtracted from total ionic weight to derive weight of water of hydration;
(c) weight of water of hydration is divided by weight per water unit (18) to derive number of water units (hydration number).
----- illustrating and validating Lewis Herrick Flint's description of hydrational potentiality
(which may in the future be referred to as Flint's Laws Of Hydration):

H (hydration number) = 23n - Z+C;
when Z = atomic number; C = valence; and n=1 for (Z+C)=0-23, n=2 for (Z+C)=23-46, n=3 for (Z+C)=46-69, and n=4 for (Z+C)=69-92.

Equation used in calculations: Hcalc = (k/(conductance^2) - AW)/18; k=517336

AN INVERSE LINEAR RELATION BETWEEN: ATOMIC-NUMBER-PLUS-VALENCE; AND HYDRATION NUMBERS DERIVED FROM IONIC WEIGHTS CALCULATED AS INVERSE-SQUARE-OF-CONDUCTANCE/MOBILITY PER LAW OF KINETIC ENERGY

by Stuart Hale Shakman, P.O. Box 382, Santa Monica, CA90406; 26 September 1996

For cations with Z = 3-19[1] -- Li+, Na+, Mg++, Al+++ and K+ -- an approximate inverse linear relation between calculated hydration numbers (H) and sums of atomic number (Z) plus valence (C) is exhibited when solute ionic weight is calculated as the inverse-square-of-conductance and adjusted to a base weight value of 85.4768 for Rb+. The subtraction of atomic weights (A.W.) from solute ionic weights yields calculated values for the weights of water of hydration, against zero[2] for Rb+; weights of water of hydration, divided by the weight of a single water unit, 18, yield calculated hydration numbers. (See Table 1).

Figure 1 plots calculated hydration numbers against respective sums of atomic number and valence. The inverse linear result shown in Figure 1 illustrates the essential foundation of a methodology first discussed by L. H. Flint in 1932[3], wherein, for the hydrated lighter ions, Z + C + H was proposed to equal 23.

As shown in Table 1 and Figure 1, calculations for H+, OH-, and the base ion, Rb+, also yield an approximate linear result at H = 0. The suggestion that relatively large conductivities of H+ and OH- indicate they are not hydrated was first made by Abegg and Bodlander[4], first calculated by Flint[3], and explained by Flint as evidencing dehydration due to the electrical stress imposed in measuring conductance[5].

1. Noggle, J.H., Physical Chemistry, 1996, p. 411.
2. Gluekauf, E., Faraday Soc., Transactions 51 1241 (1955).
3. Flint, L. H., J. Wash. Acad. of Sci. 22, 97-119, 211-217 & 233-237 (1932).
4. Abegg and Bodlander, Zeit. f. Anorg. Chem. 454-499 (1899).
5. Flint, L.H., Dissenting Ape, Dahlia Street, New York, 1973.2

Table 1: Inverse square of limiting ionic conductance (l) adjusted to base weight of 85.47 for Rb+; minus respective atomic weights (A.W.), equals weight of water of hydration, divided by 18 equals calculated hydration number (H).

ION C Z	A.W.	Cond.	517336/(Cond.^{2}) =IONIC WEIGHT	-A.W. =WATER WEIGHT	/18 =Hcalc	Hcalc+Z+C
Rb +1 37	85.4678	77.8	85.47(BASE)	.00	.00	
Li +1 3	6.941	38.69	345.59	338.65	18.81	22.81
Na +1 11	22.989768	50.11	206.02	183.03	10.17	22.17
Mg +2 12	24.3050	53.06	183.75	159.45	8.86	22.86
Al +3 13	26.981539	63	130.34	103.36	5.74	21.74
K +1 19	39.0983	73.50	95.76	56.66	3.15	23.15
H +1 1	1.0078	349.8	4.23	3.22	.18	
OH -1 9	17.00734	197.6	13.25	-3.76	-.21	

```
Figure I.  Correlation of calculated hydration numbers (H) with sums of atomic numbers and valences (Z+C)
Z+C:
38-    *      0 (Rb+ = BASE)
37-    |
36-
35-    |
34-
33-    |
32-
31-    |
30-
29-    |
28-
27-    |
26-
25-    |
24-
23-    |
22-
21-    |
20-            *    3.15 (K+)
19-    |
18-
17-    |
16-                 *    (Al+++ = 5.74)
15-    |
14-                     *    (Mg++ = 8.86)
13-    |
12-                        *    (Na+ = 10.17)
11-    |
10-
09-    |
08-    *    (OH- = -.21)
07-    |
06-
05-    |                                              *    18.81 (Li+)
04-                                                 *    18.81 (Li+)
03-    |
02-    *    (H+ = .18)
01-    |
00----------------------------------------------------------------------------
     |  |  |  |  |  |  |  |  |  |  |  |  |  |  |  |  |  |  |  |  |  |  |  |
H:   0  1  2  3  4  5  6  7  8  9  10 11 12 13 14 15 16 17 18 19 20 21 22 23
```

*The above article was submitted to Nature, 26 Sept. 1996 as proposed scientific correspondence; registered by Nature as SXA011; and subsequently rejected without further explanation

Associated Correspondence:

P.O. Box 382
Santa Monica, CA90406-0382
(310)453-7707; ae735@lafn.org
26 September 1996

Editor, Nature
1234 National Press Building
Washington, D.C. 20045

Dear Nature Editor,

Thank you for having printed my previous submission, "Heliocentric tangents", in your "Correspondence" section (Nature 338, 456, 1989).

Enclosed is a short paper intended for consideration for your "Scientific correspondence" section, entitled "AN INVERSE LINEAR RELATION BETWEEN: ATOMIC-NUMBER-PLUS-VALENCE; AND HYDRATION NUMBERS DERIVED FROM IONIC WEIGHTS CALCULATED AS INVERSE-SQUARE-OF-CONDUCTANCE/MOBILITY PER LAW OF KINETIC ENERGY".

This submission is thought to be of possible interest to Nature readers due to the generally-accepted importance of gaining an improved understanding of aqueous solution phenomena, as reflected in former Nature editor John Maddox's expression of frustration over our inability to calculate such phenomena (Nature 364, 483, 1993).

Thank you for your kind consideration.

Sincerely yours,

Stuart Hale Shakman

enclosure

Table: Calculation of hydration numbers (Hcalc) from electrical conductance (EC) & atomic weight (AW)

as Hcalc=((k/EC-square)-AW)/18; k=517322.9186 when base is Rb+ with Hcalc=0. Copyr.1999 SHShakman

ION	Z	C	AW	EC	IW calc	WW	H calc	Z+C	(Z+C+H)/23
Rb (BASE)	37	1	85.4678	77.8	85.4678	0	0	38	
Li	3	1	6.941	38.66	346.129	339.188	18.84378	4	0.993208
Be	4	2	9.012182	45	255.4681	246.4559	13.692	6	0.856174
Na	11	1	22.98977	50.08	206.2686	183.2788	10.18216	12	0.964442
Mg	12	2	24.305	53.06	183.7499	159.4449	8.858053	14	0.993828
Al	13	3	26.98154	63	130.3409	103.3593	5.742185	16	0.945312
K	19	1	39.0983	73.48	95.81278	56.71448	3.150804	20	1.006557

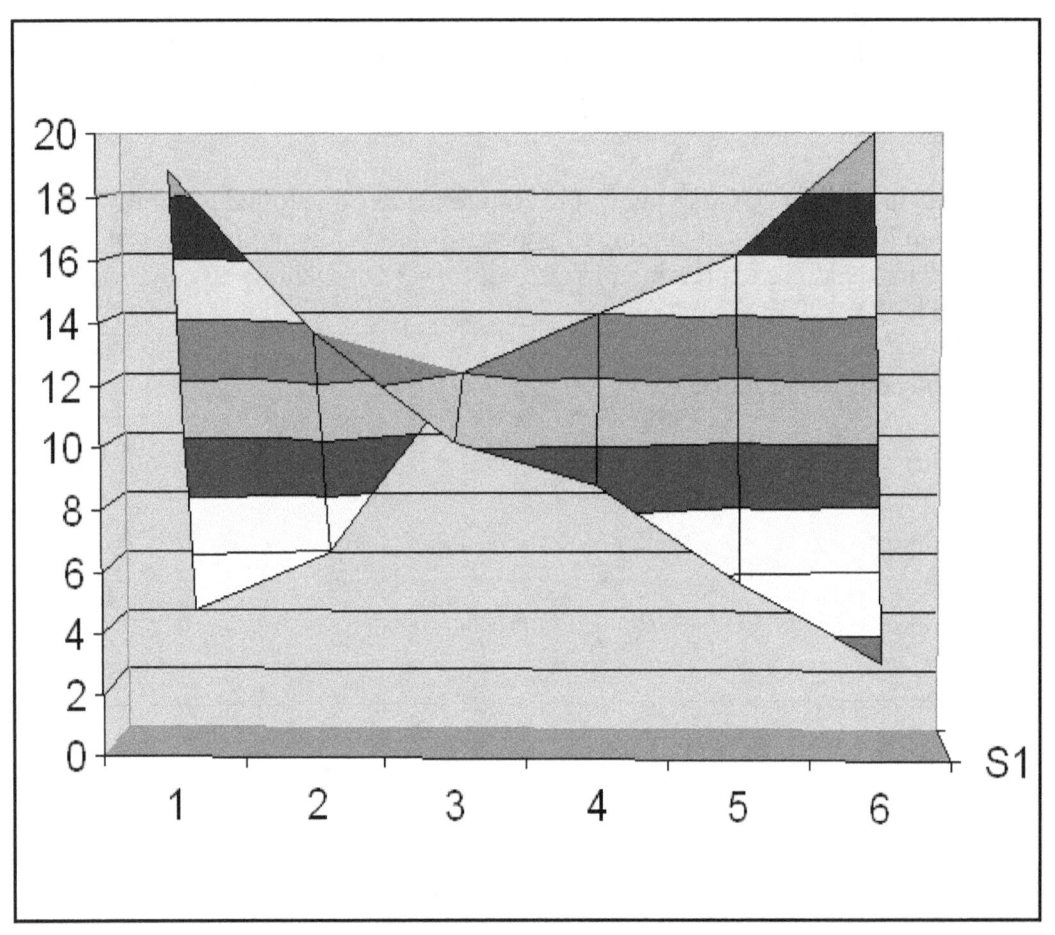

Primacy of <u>Atomic-Number-Plus-Valence</u> versus <u>Atomic-Weight</u>

The realization of an inverse integral relationship between atomic and hydration numbers respectively, in combination with the observation of an improved order of agreement when atomic numbers are adjusted per unit valence, led to Flint's adoption of use of Z+C as measure of relative mass, in preference to Atomic Weight.

The below table (AW versus Z versus Z+C) exhibits calculations of hydration numbers using each of three ions as the base ion (H+, Rb+ or La+3) in each of three sets of calculations, inputting Atomic Weights, versus Atomic Numbers, versus Atomic Numbers + Valence, respectively. Thus, within the first set of calculations, that inputting Atomic Weights, for the middle column using Rb+ as the base ion, the result is that which is also illustrated in the paper <u>"Direct Calculation of Hydration Numbers for the Lighter Ions"</u>; the other columns merely substitute H+ and La+3 resp. as the base ion in place of Rb+; and the other sets of calculations substitute Z or Z+C resp. for AW:

As is shown, the order of agreement between calculations using all three base ions is clearly preferable using Z+C versus either Z or AW:

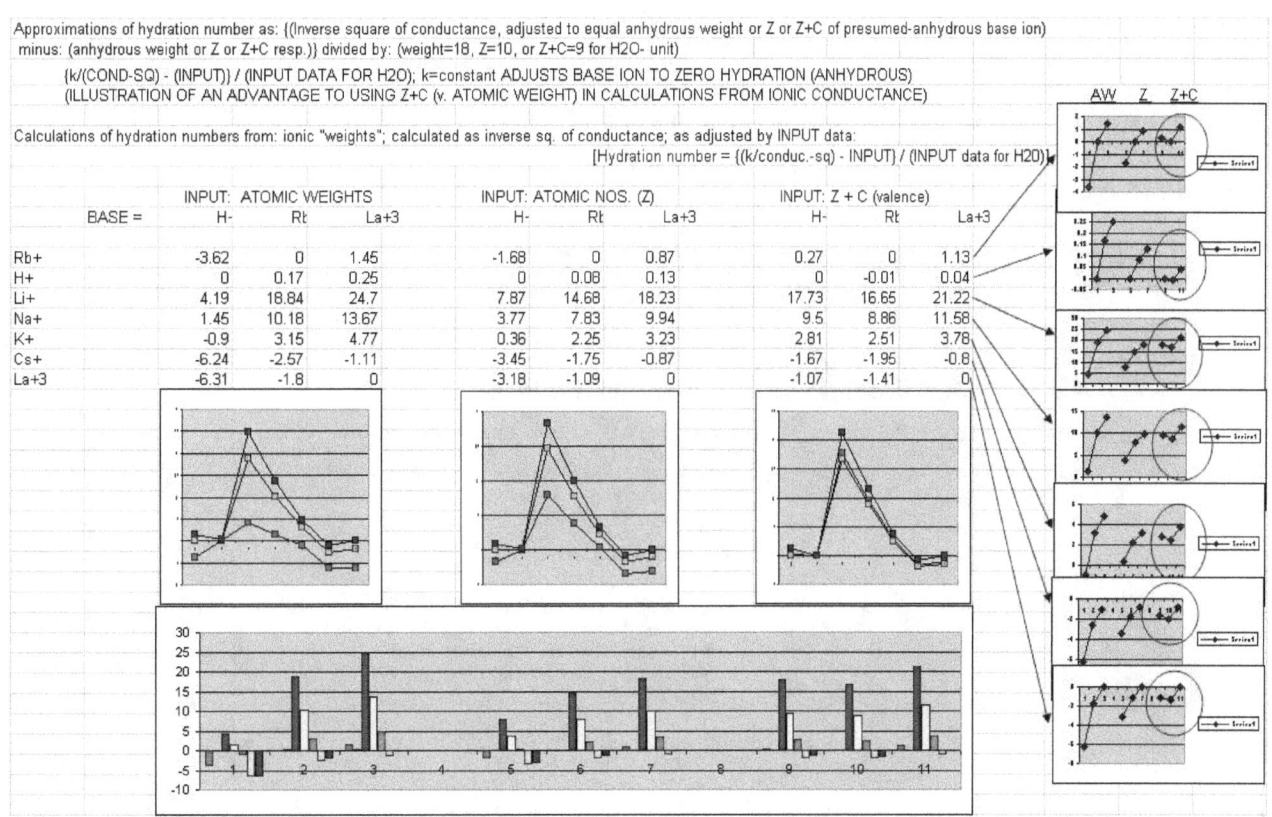

Approximations of hydration number as: {(Inverse square of conductance, adjusted to equal anhydrous weight or Z or Z+C of presumed-anhydrous base ion) minus: (anhydrous weight or Z or Z+C resp.)} divided by: (weight=18, Z=10, or Z+C=9 for H2O- unit)

{k/(COND-SQ) - (INPUT)} / (INPUT DATA FOR H2O); k=constant ADJUSTS BASE ION TO ZERO HYDRATION (ANHYDROUS)
(ILLUSTRATION OF AN ADVANTAGE TO USING Z+C (v. ATOMIC WEIGHT) IN CALCULATIONS FROM IONIC CONDUCTANCE)

Calculations of hydration numbers from: ionic "weights"; calculated as inverse sq. of conductance; as adjusted by INPUT data:
[Hydration number = {(k/conduc.-sq) - INPUT} / (INPUT data for H2O)]

BASE =	INPUT: ATOMIC WEIGHTS			INPUT: ATOMIC NOS. (Z)			INPUT: Z + C (valence)		
	H-	Rb	La+3	H-	Rb	La+3	H-	Rb	La+3
Rb+	-3.62	0	1.45	-1.68	0	0.87	0.27	0	1.13
H+	0	0.17	0.25	0	0.08	0.13	0	-0.01	0.04
Li+	4.19	18.84	24.7	7.87	14.68	18.23	17.73	16.65	21.22
Na+	1.45	10.18	13.67	3.77	7.83	9.94	9.5	8.86	11.58
K+	-0.9	3.15	4.77	0.36	2.25	3.23	2.81	2.51	3.78
Cs+	-6.24	-2.57	-1.11	-3.45	-1.75	-0.87	-1.67	-1.95	-0.8
La+3	-6.31	-1.8	0	-3.18	-1.09	0	-1.07	-1.41	0

PRESENTATIONS 1985-1996

Archive:Atomic Weight paper 9 Dec. 1985

Shakman, S.H., ATOMIC WEIGHT   1.   *As distrib. to Hansen*
Abelson
D'Amico

ATOMIC WEIGHT

or

ON THE RELATION BETWEEN ATOMIC NUMBERS AND ATOMIC WEIGHTS

Stuart Hale Shakman

9 December, 1985

Part A.   The Weight of Hydrogen Gas, based on a study of

interdiffusion of gases.

Part B.   Defense of Part A.

Part C.   Potential Implications and Speculations.

Please note that Part A. is self-contained (includes notes and
references) and could be published separate from the remainder;
however, it would be preferred that all three parts be published
together.

Phone: (415)673-6546

Shakman, S.H., ATOMIC WEIGHT   2.   (Part A., page A-1.)

Part A. The Weight of Hydrogen Gas, based on a study of
             interdiffusion of gases.

Abstract: A study of interdiffusion of gases supports the hypothesis of a direct correlation between atomic numbers of the elements
and their respective atomic weights. In 47 of 48 groups of interdiffusing gas-pairs, comparisons of observed and theoretical rates
of interdiffusion support a value of 4 for the weight of H_2 gas, in
preference to a value of 2, on a scale of 32 for the weight of O_2
gas. Results in the 48th group may be considered ambiguous as discussed below in "Analysis".[1]

Background: The elements hydrogen, with an atomic number of one,
and oxygen, with atomic number 8, are conventionally assigned weight
values 1.00794 ± 7 and 15.9994 respectively (ref.h). Thus H_2 and O_2
gases are considered to have molecular weights of approximately 2
and 32 respectively.

An alternate scale with atomic weights proportional to atomic
numbers was utilized by Lewis H. Flint in 1932 in calculations
within a study of hydrated ions in solution (ref.i), which scale
prescribed a value of 4 for the weight of H_2 gas relative to 32 for
O_2 gas. Subsequently in 1964 Flint labeled this proportional
scale "Corollary One" of his proposed "Description Of Hydrational
Potentiality", stating that the weight of an atom in its neutral
state is double its atomic number $\angle W_n$=2 A.N., when W_n=the weight
of the involved unit in the neutral state$\angle7$ (ref. j, p. 21). Flint
discussed his work as a validation of "Prout's hypothesis" (ref. j,
Chapters 2 & 11), after William Prout who in 1815-16 had hypothesized that the absolute weights of the elements are whole-number
multiples of the weight of hydrogen (ref. k, p. 40).[2]

Theoretical: This study rests on knowledge of a relationship between

Shakman, S.H., ATOMIC WEIGHT   3.   (Part A., page A-2.)

diffusion and weight, as reflected in Graham's law of diffusion (after
Sir Thomas Graham, 1805-1869), and efforts by Flint to extend such
knowledge through development of an expression of interdiffusion
as a function of the weights of the involved gases (ref. j, Chapter 11).

  Graham's law of diffusion describes the relative mobility (M),
or relative rate of diffusion, of a gas as inversely proportional to
the square root of its weight (W) $[M = 1/\sqrt{W}]$. Flint proposed to
describe the interdiffusion of two gases (A-B) as directly proportional
to the difference between their mobilities divided by the difference
in their weights. Flint's formulation is not used in this study as
it cannot calculate for gases of equal weights.

  As an approximation to the behavior of interdiffusing gas-pairs,
this study describes the relative rate of interdiffusion of a given
gas-pair (A-B) as the square root of: The sum of mobilities of the
involved gases divided by the product of their weights
$[I = \sqrt{(M_A + M_B)/(W_A W_B)}]$, which equasion can be expressed solely in
terms of weight:

$$I = \frac{\sqrt{W_A}\sqrt{W_B} + W_B\sqrt{W_A}}{W_A W_B}$$

Description of Exhibit A: Exhibit A lists 48 groups of gas-pairs,
each containing two or more gas-pairs (A-B); 47 of these groups consist of a single common gas ("A") interdiffusing with H_2 and at least
one other gas ("B"). The 48th group consists of self-diffusion data
for H_2 and He gases.

  For each gas-pair listed in Exhibit A, an observed diffusion
coefficient (D_O), references (refs), and relative ratio(s) (RR, RR',
as described below) are also shown. Where multiple values for D_O are
found in references, that listed is an average of high and low values.

  RR: In each of the 48 groups, the ratio of observed to
calculated values (D_O/I) for the second gas-pair is adjusted to a
base value of 1.000 by dividing it by itself, and ratios (D_O/I) for

Shakman, S.H., ATOMIC WEIGHT   4.   (Part A., page A-3.)

all other gas-pairs within that group are adjusted proportionately
through division by the same factor $\underline{/}RR=(D_O/I)_i \div (D_O/I)_2$, where i=1
denotes the first gas-pair in each group, i=2 the second, etc$\underline{.}\underline{/}$.

  Example: For $CO-H_2$, $RR = \dfrac{D_O/I \text{ for } CO-H_2}{D_O/I \text{ for } CO-CO} = \dfrac{.651/.07843}{.183/.02196} \approx \dfrac{8.30}{8.33} = .996$.

All $\underline{RR}$ values (☉,·) are derived from theoretical (I) values
that are based on the assumption that the weight (W) of each involved
gas is proportional to, i.e. equal to double, the sum of atomic numbers
(ref.h) of its component atoms (Thus, $O_2=32$, $N_2=28$, and $H_2=4$).

  $\underline{RR'}$: For each gas-pair involving H_2 gas, an alternate value,
designated "$\underline{RR'}$" (X), is derived from a theoretical (I') value that
is based on the alternate assumption that H_2 gas weighs 2 while
gases not containing hydrogen continue to be assigned weight values
equal to double the sums of atomic numbers of component atoms (Thus $O_2=32$,
$N_2=28$, and $H_2=2$, values approximating conventionally accepted values).
Note that "I'" and "$\underline{RR'}$" are calculated in accordance with above
equations for "I" and "$\underline{RR}$" respectively.

Analysis: In 47 of the 48 groups listed, RR (☉) values for gas-pairs
involving H_2 gas are closer to the base value of 1.000 than are
corresponding RR' (X) values. Note also the variation in identities
of the "B" gases in the second gas-pairs of these 47 groups: 11 are
He, 10 are N_2, 18 are air ($W_{air}=28.8$), and 8 are (five) other gases.

  Results in the 48th group may be considered ambiguous: In the
group in which Rn is the common ("A")gas, RR (☉) is further from
the base value of 1.000 than is RR' (X); however, the order of
agreement between the RR (☉) value and RR (·) values shown for the
three gas-pairs other than the second gas-pair is preferable to that
between these other three RR (·) values and the RR' (X) value.

Shakman, S.H., ATOMIC WEIGHT   5.   (Part A., page A-4.)

<u>Conclusion</u>: Results of this study constitute evidence that on a
scale of 32 for the weight of O_2 gas, H_2 gas weighs 4. This evidence
may be taken as argument in support of the validity of Corollary One
of Flint's proposed Description Of Hydrational Potentiality.[3]

#

Notes: 1. Full study involved a total of 55 groups of gas-pairs; seven of the
55 groups were excluded from this 11 Aug. paper in that four consisted of one gas-pair only, each involving H_2 gas but with no second
gas-pair for comparison, and the remaining three did not contain a
gas-pair involving H_2 gas (full report, calculation tables, and de-
tailed notes available on request).

2. It is beyond the scope of this study to attempt to reconcile the
conventionally-accepted weight value for H_2 gas (2.016) with that
proposed by Flint and supported by this study (4); however, it may
be noted that a possible reconciliation may be derived from Flint's
"Corollary Two" which prescribes an orderly weight change with ionization $\sqrt{W_a=2(A.N.\pm C)}$, when W_a=weight as an anhydrous ion and C=
valence⧸ (ref. j, p. 22). Through the use of Corollary Two, Flint
proposed to express mathematically how compounds utilized in conventional atomic weight determinations might be comprised of elemental ions with weights other than those prescribed for isolated
neutral elements (ref. j, Chapter 18). In accordance with this
corollary, Flint projected a value of 18 for the weight of water
based on the presumption of water's existence as the negatively
charged "H_2O^-" ion (ref. j, p. 25) in contrast to a value of 20
which would be prescribed for a neutral H_2O molecule; this projected value of 18 approximates the conventionally-accepted weight
value for water, which latter value figures prominently in conventional determinations of relative atomic weight values for the
elements hydrogen and oxygen (ref. l, p. 11-31). The hypothesis
that an entity may weigh other than the sum of weights of constituent
atoms may be viewed as an extension of the suggestion made by C.
Marignac in 1860, in comments supporting Prout's hypothesis, that
a grouping of primordial atoms in the form of a (more complex)
chemical atom might weigh other than the sum of weights of constituent
primordial atoms (ref. k, p. 58); similarly, in 1921 F.W. Aston
proposed the concept of a "packing" effect which might allow for
an atomic nucleus to weigh less than the sum of weights of constituent
"charges" (ref. k, p. 22).

3. As proposed by Flint (ref. j, p.21-22), hydrational potentiality may
be described by the formula: <u>$H=23N-(A.N.\pm C)$</u> and corollaries, in which
<u>H</u>   = the number of H_2O^- units of weight 18 (see Note 2 above)
which may maximally be held by an ion in solution;
<u>A.N.</u>= the atomic number of the element in the case of an element
ion, or in the case of a molecular ion A.N.=the sum of
atomic numbers of component atoms;
<u>C</u>   = the valence of the ion, this number to be added if positive,
subtracted if negative; and
<u>N</u>   = an integer representing one of four periods: For atomic
numbers 0-22, N=1; 23-45, N=2; 46-68, N=3; 69-92, N=4.
Four corollaries delimit categories of weight of the involved units:

Shakman, S.H., ATOMIC WEIGHT   6.   (Part A., page A-5.)

Corollary One- $W_n = 2$ A.N., when W_n=weight in the neutral state;
Corollary Two- $\overline{W_a = 2(A.N. \pm C)}$, When W_a=weight as an anhydrous ion;
Corollary Three- $\overline{W_c = (W_n + W_a)/2}$, when W_c=combining weight; and
Corollary Four- $\overline{W_h = W_a + 18H}$, when W_h=maximally hydrated weight.

Flint characterized his description as a mathematical extension of Abegg and Bodländer's recognition of an inverse relationship between ionic weight and hydration (ref. m), which extension was derived from use of Graham's law of diffusion in conjunction with Kohlrausch ionic conductance data for K^+ and Na^+ ions (ref.i,p.99, and ref.j,p.16-17). Accordingly, it may be noted that the ratio (1.554:1) between observed conductivities of 40.4 for K^+ and 26 for Na^+ at $0^\circ C$ (ref. n) closely approximates the ratio (1.537:1) between theoretical mobilities of these two ions as calculated in the below example.

Example: For the K^+ ion, with A.N.=19 (as per ref. h), N=1, C=+1, and $H = \underline{/}23(1)-(19+1)\underline{/}=3$ (as per above formulaic representation of Flint's description); $W_a = \underline{/}2(19+1)\underline{/}=40$ (as per Corollary Two; and $W_h = \underline{/}40+(18\cdot3)\underline{/}=94$ (as per Corollary Four).

Similarly, for the Na^+ ion, with A.N.=11 (ref.h), N=1, C=+1, and $H = \underline{/}23(1)-(11+1)\underline{/}=11$; $W_a = \underline{/}2(11+1)\underline{/}=24$; and $W_h = \underline{/}24+(18\cdot11)\underline{/}=222$.

Theoretical relative mobilities for fully hydrated K^+ and Na^+ ions may be calculated in accord with Graham's law as the inverse of the square roots of their respective weights (W_h), or $1/\sqrt{94} = .1031416$ and $1/\sqrt{222} = .0671155$ respectively, for a relative ratio of 1.537:1.

References:

a. INTERNATIONAL CRITICAL TABLES, Vol. 5. McGraw Hill, 1929, p. 62-63.
b. AMERICAN INSTITUTE OF PHYSICS HANDBOOK, McGraw Hill, 1957, p. 2-212, 2-213.
c. Jost, W., DIFFUSION IN SOLIDS, LIQUIDS, GASES. Academic Press, Inc., 1960, p. 409, 412-413, 430.
d. Chapman, S., and Cowling, T.G., THE MATHEMATICAL THEORY OF NON-UNIFORM GASES, Cambridge University Press, 1970, p. 263, 266-7.
e. Hirschfelder, J.O., Bird, R.B., and Spotz, E.L., CHEM. REV. 44, 209, 226 (1949).
f. Grew, K.E., and Gibbs, T.L., THERMAL DIFFUSION IN GASES. Cambridge University Press, 1952, p. 71.
g. Hirschfelder, J.O., Curtiss, C.F., and Bird, R.B., MOLECULAR THEORY OF GASES AND LIQUIDS. John Wiley & Sons, Inc., 1954, p. 579, 601.
h. CRC HANDBOOK OF CHEMISTRY AND PHYSICS, 1984-85. (Atomic Numbers:
   H - 1   C - 6   O - 8   S -16   K -19   Cd-48   Xe-54   Rn-86
   He- 2   N - 7   Na-11   A -18   Kr-36   I -53   Hg-80   ).
i. Flint, L.H., J. WASH. ACAD. SCI., 22, 97-119 & 22, 211-217 (1932).
j. Flint, L.H., BEHAVIOR PATTERNS OF HYDRATION. Institute for the Advancement of Science and Culture, New Delhi, 1964.
k. Prout, W. (1815-16), J.S. Stas (1860) and C. Marignac (1860), PROUT'S HYPOTHESIS. L.C.# QD 463.P7.
l. Clarke, F.W., MEMOIRS N.A.S., Vol. 16, No. 3 (1922).
m. Abegg,R. & Bodländer, Die Elektroaffinität, ein neues Prinzip der chemischen Systematik. Z.Anorg.Chem.20(1899),453, per. ref.j, p.20.
n. CRC HANDBOOK OF CHEMISTRY AND PHYSICS, 48th Edition (1967-69).

Shakman, S.H., ATOMIC WEIGHT   7.   (Part A., page A-6.)

Shakman, S.H., ATOMIC WEIGHT   8.   (Part B., page B-1.)

Part B. Defense of Part A.

It may be noted that the element sodium is involved in Part A in two contexts: (1) as a vapor (Na) interdiffusing with N_2 gas as shown in Exhibit A on p.7 and (2) as a solute ion (Na^+) whose equivalent conductivity is contrasted with that of the potassium ion (K^+) as shown in Note 3. on p. 6 (Part A., page A-5.).

It is further noted that in this first context the diffusion coefficient involving sodium (D_o for N_2-Na=20.4) is the largest value listed in Exhibit A, yielding a relatively large "RR" value (103.177) that falls far outside the range of all other RR values (0.6-2.2); whereas in the second context the conductivity of Na^+ (26) relative to that of K^+ (40.4) is utilized as the exemplary base for derivation of Flint's proposed description.

This defense seeks to justify (1) disregarding the measurement involving sodium in the first context (interdiffusion) while supporting as satisfactory the general order of agreement between calculated and other observed (interdiffusion) values in Part A; and (2) embracing the measurement involving sodium in the second context (equivalent conductivity) while demonstrating the potential utility of Flint's description and corollaries.

(1)   One interdiffusion measurement only involving sodium was found in references a. - g. listed on p. 6 (Part A, p.A-5) as shown in Exhibit A (D_o for N_2-Na=20.4). This relatively large D_o value, a value 30 times that of N_2-H_2 (D_o=.676) would appear to be either (a) an error or (b) indicative that a very small particle is involved - - one far smaller than H_2 gas. This latter possibility might be illustrated by noting in Exhibit A that the observed diffusion coefficient for N_2-H_2 is ten times as great as that for N_2-I_2 (D_o=.068), while the prescribed

Shakman, S.H., ATOMIC WEIGHT   9.   (Part B., page B-2.)

weight (W_n=2 A.N., as per Flint's Corollary One) for H_2 (4) is about one-fiftieth that for I_2 (212).

Regardless of the disposition of the question as to why the published D_O for N_2-Na is anomolous, its very solitary existence and that of its corresponding "RR" value reflect favorably on the order of agreement existing between the remainder of observed and calculated interdiffusion values.

In this connection it may be noted in Exhibit B-1 that calculated diffusion coefficient values (D_C) obtained using the equasion for "I" shown on p. 3 (Part A, page A-2), adjusted to the base of D_C for N_2-O_2=.181 $\left[D_C = I_{A-B} \cdot \dfrac{(D_C \text{ for N2-O2}=.181)}{(I \text{ for N2-O2}=.0202042)} = I_{A-B} \cdot 8.9585 \right]$, reflect a generally equally satisfactory order of agreement with observed values as do the results obtained by Hirschfelder, etal. in ref. g., p. 579(D_C"). Be advised that Hirschfelder's calculations utilize experimental viscosity data and conventionally-accepted atomic weight values, whereas those of subject paper utilize atomic numbers exclusively and thus are more direct. Note also, for gas-pairs involving H_2 gas, the preferability of D_C values listed in Exhibit B-1, based on the assumption that H_2 gas weighs 4 as per p. 3 (Part A,p.A-2) to D_C' values, based on the assumption that H_2 weighs 2 as per p.3-4 (Part A,p.A-2,A-3) as evidenced in the proximity of R and R' values respectively to 1.0.

(2) References b, h, and n on p.6 (Part A, p.A-5) contain several conductivity measurements involving both sodium and potassium. Hence a measure of support for the integrity of the ratio between conductivities of Na^+ and K^+ ions as used in Note 3. of Part A may be found in comparable data at elevated temperatures as listed in Exhibit B-2. Although the number of water molecules attracted to a given ion apparently decreases as temperature increases [Nernst, W., THEORETICAL CHEMISTRY (1916) p. 397.], it may be noted that the actual ratio between conductivities of K^+ and Na^+ ions at elevated temperatures

Shakman, S.H., ATOMIC WEIGHT   10.   (Part B., page B-3.)

(Column 5 of Exhibit B-2) remains within one-tenth of the prescribed ratio of 1.554:1 (as per Note 3. of Part A) through 50°C.

Furthermore it may be noted in Exhibit B-3 that at 25°C a majority of 47 observed conductivity values for positively-charged inorganic ions listed in ref. h, p. D-167 and D-168, are within the range of 1.0 $\pm$ .1 of values prescribed by Flint's description and corollaries, utilizing as the base value that of Na^+=50.08 and under one of two assumptions - either full or zero hydration. Note that 13 of these observed values fall into the relatively narrow range of 1.045 $\pm$ .020 of prescribed values.

‡

Shakman, S.H., ATOMIC WEIGHT   11.   (Part B., page B-4.)

Exhibit B-1: Comparison of Observed Diffusion Coefficients (D_O) at $0^\circ C$
with Three Calculations (D_C, D_C', and D_C'' as per below Key)

Key: A - B = gas-pair.

D_O = observed diffusion coefficient (rate of inter-diffusion), from Exhibit A, p. 7 (Part A, p.A-6.).

D_C = (I for gas-pair A-B) $\cdot \dfrac{(D_C \text{ for } N_2\text{-}O_2 = .181)}{(I \text{ for } N_2\text{-}O_2 = .0202042)}$

= $I_{A\text{-}B} \cdot 8.9585$; $I = \dfrac{\sqrt{W_A}\sqrt{W_B} + W_B\sqrt{W_A}}{W_A W_B}$ (H_2 weighs 4).

R = ratio: D_O/D_C.

D_C' = calculation   (as per description of D_C above; alternate assumption – H_2 weighs 2).

R' = ratio: D_O/D_C'.

D_C'' = calculation per ref. g., p. 579.

R" = ratio: D_O/D_C''.

A -	B	D_O	D_C	R=D_O/D_C	D_C'	R'=D_O/D_C'	D_C''	R"=D_O/D_C''
CH_4	H_2	.625	.852	(.734)	1.497	(.431)	.607	(1.030)
	CO_2	.153	.185	(.827)			.138	(1.109)
N_2	H_2	.674	.703	(.962)	1.133	(.597)	.656	(1.027
	CO	.192	.197	(.975)			.174	(1.103)
	O_2	.181	.181	(1.000)	BASE		.175	(1.034)
	CO_2	.144	.149	(.966)			.130	(1.108)
CO	H_2	.651	.703	(.926)	1.133	(.575)	.661	(.985)
	N_2	.192	.197	(.975)			.174	(1.103)
	O_2	.185	.181	(1.030)			.175	(1.057)
	CO_2	.137	.149	(.919)			.128	(1.070)
O_2	H_2	.697	.651	(1.071)	1.053	(.662)	.689	(1.012)
	N_2	.181	.181	(1.000)			.175	(1.034)
	CO	.185	.181	(1.030)			.175	(1.057)
	CO_2	.139	.137	(1.011)			.128	(1.086)
CO_2	H_2	.550	.545	(1.002)	.885	(.617)	.544	(1.011)
	N_2	.144	.149	(.966)			.130	(1.108)
	CO	.137	.149	(.919)			.128	(1.070)
	O_2	.139	.137	(1.011)			.128	(1.086)
N_2O	H_2	.535	.545	(.982)	.885	(.605)	.552	(.969)
	CO_2	.096	.112	(.866)			.092	(1.043)
A	He	.641	.610	(1.051)			.653	(.982)

Shakman, S.H., ATOMIC WEIGHT   12.   (Part B., page B-5.)

Exhibit B-2: Comparison of Equivalent Ionic Conductivities for K^+
and Na^+ Ions at Various Temperatures.

Key: 1. Ref.   = reference.

2 Temp.   = temperature (°C).

3. $\triangle_o$, K^+ = Equivalent Ionic Conductivity for K^+ ion.

4. $\triangle_o$, Na^+ =   "   "   "   " Na^+ " .

5. Ratio   = Ratio: $\triangle_o$, K^+ / $\triangle_o$, Na^+ .

1. Ref.	2. Temp.	3. $\triangle_o$, K^+	4. $\triangle_o$, Na^+	5. Ratio
n	0°C	40.4	26	1.554
b	15	59.66	39.75	1.501
n	18	64.6	43.5	1.485
b	25	73.50	50.11	1.467
h	25	73.48	50.08	1.467
n	25	74.5	50.9	1.464
b	35	88.21	61.53	1.434
n	50	115	82	1.402
n	75	159	116	1.371
n	100	206	155	1.329
n	128	263	203	1.296
n	156	317	249	1.273

Prescribed (see Part A, Note 3.) 1.537

Shakman, S.H., ATOMIC WEIGHT   13.   (Part B., page B-6.)

Exhibit B-3:   Comparison of Observed Equivalent Conductivity Values
at 25°C ($\triangle_o$) with Calculated Values ($\triangle_c$).

Key: 1. Ion   = inorganic positively-charged ion listed
in ref. h, p. D-167 and D-168.

2. $\triangle_o$ = observed equivalent conductivity per ref. h.

3. A.N.   = atomic number.

4. (h) or (a) = assumption: hydrated (h) or anhydrous (a).

5. W_t = theoretical weight of solute ion (either
W_h or W_a as indicated by assumption in Column 4.

6. M_t = theoretical relative mobility of solute ion
= $1/\sqrt{W_t}$.

7. $\triangle_c$ = calculated equivalent ionic conductivity
value (to the base Na^+=50.08)

= (M_t for Ion) $\cdot$ $\dfrac{(\triangle_c \text{ for } Na^+ = 50.08)}{(M_t \text{ for } Na^+ = .06712)}$

= $M_t \cdot 746.126$.

8. Ratio   = $\triangle_o \div \triangle_c$ (Column 2 $\div$ Column 7) .

1. Ion	2. $\triangle_o$	3. A.N.	4. (h,a)	5. W_t	6. M_t	7. $\triangle_c$	8. Ratio
Al^{3+}	61	13	(h)	158	.07956	59.4	1.028
Ba^{2+}	63.6	56	(a)	116	.09285	69.3	.918
Be^{2+}	45	4	(h)	318	.05608	41.8	1.075
Ce^{3+}	69.8	58	(a)	122	.09054	67.6	1.033
Cs^+	77.2	55	(a)	112	.09449	70.5	1.095
Dy^{3+}	65.6	66	(a)	138	.08513	63.5	1.033
Er^{3+}	65.9	68	(a)	142	.08392	62.6	1.052
Eu^{3+}	67.8	63	(a)	132	.08704	64.9	1.044
Gd^{3+}	67.3	64	(a)	134	.08639	64.5	1.044
H^+	349.65	1	(a)	4	.50000	373.1	.937
Hg^{2+}	63.6	80	(a)	164	.07809	58.2	1.090
Ho^{3+}	66.3	67	(a)	140	.08452	63.1	1.051
K^+	73.48	19	(h)	94	.10314	77.0	.954
La^{3+}	69.7	57	(a)	120	.09129	68.1	1.078
Li^+	38.66	3	(h)	350	.05345	39.9	.969
Mg^{2+}	53.0	12	(h)	190	.07255	54.1	.980
Na^+	50.08	11	(h)	222	.06712	50.08	1.000 (BASE)
Nd^{3+}	69.4	60	(a)	126	.08909	66.5	1.044
Pr^{3+}	69.5	59	(a)	124	.08980	67.0	1.037
Rb^+	77.8	37	(a)	76	.11471	85.6	.909
Sm^{3+}	68.5	62	(a)	130	.08771	65.4	1.046
Sr^{2+}	59.4	38	(h)	188	.07293	54.4	1.091
Tm^{3+}	65.4	69	(a)	144	.08333	62.2	1.052
Y^{3+}	62	39	(h)	156	.08006	59.7	1.037
Yb^{3+}	65.6	70	(a)	146	.08276	61.7	1.063

Shakman, S.H., ATOMIC WEIGHT   14.   (Part C., page C-1.)

Part C. Potential Implications and Speculations

In the final paragraph of his "Relativistic Theory of the Non-Symmetric Field" (Appendix II of the Fifth Edition of THE MEANING OF RELATIVITY, published in 1956), Albert Einstein discussed the inability of a continuum theory to describe reality, concluding that this "must lead to an attempt to find a purely algebraic theory for the description of reality. But nobody knows how to obtain the basis of such a theory."

As discussed in Part A of this paper, more than two decades earlier Lewis H. Flint had initially proposed an algebraic system purporting to explain the behavior of solute ions in aqueous solutions, which behavior is of fundamental and common concern within the fields of physics, chemistry, and biology. Such a system, if found to be valid, would constitute a significant measure of unification among the involved sciences, and thus might logically constitute some portion of, or be intimately related to, a more general algebraic system underlying these sciences.

Were such a comprehensive system to exist, it would likely be reflected to some extent in observable phenomena, as well as reflect on its own mathematical origins. It is within this highly speculative context that a body of curious coincidences is herein introduced:

(1) Significance of the number 23 relative to numbers of water molecules in

   (a) Flint's works (Exhibit C-1),

   (b) in Structure I gas hydrates (Exhibit C-2), and

   (c) as indicated in Berecz and Achs-Balla's graph of maximum hydration numbers (Exhibit C-3);

Shakman, S.H., ATOMIC WEIGHT   15.   (Part C., page C-2.)

(2) Geometrical similarity between

   (a) Flint's projected carotin structure (Exhibit C-4),

   (b) observed structure of photo-synthetic membrane (Exhibit
       C-5), and

   (c) micrograph of earthworm hemoglobin (Exhibit C-6);

(3) Significance of the number 24 in

   (a) Flint's matrix (Exhibit C-7),

   (b) geometry of four-dimensional balls (Exhibit C-8),

   (c) Russian literature on sub-atomic particles (Exhibit C-9), and

   (d) Einstein's theory on planetary orbits (Exhibit C-10);

Note also the significance of the number 6, a factor of 24,
in Exhibits C-8 and C-12, as well as in the geometry of
Exhibits C-4, C-5, and C-6; the factor 12 in Exhibits C-8 and
C-10; and the multiple 48 in Exhibit C-10.

(4) Measure of symmetry exhibited by

   (a) superimposition of Mendeleev's periodicity on Flint's
       matrix as shown in a proposed combined periodic table of
       elements (Exhibit C-11), and

   (b) prime numbers (plus the number one) within Flint's matrix,
       as evident in the superimposition of the second half of
       Flint's matrix in reverse order beginning with the number
       89, on the first half beginning with the number one
       (Exhibit C-12); note also how this symmetry involves patterns
       of 6 as highlighted by the arrangement in six columns;

(5) The first instance of prime numbers separated by a gap
     exceeding five digits, that involving prime numbers 89 and
     97 (as highlighted in Exhibit C-13),

   (a) encompasses the number 92, the largest of atomic numbers
       of the naturally occurring elements
       and termination point of Flint's four periods, and

   (b) coincides with the beginning of the "Actinides" of

Shakman, S.H., ATOMIC WEIGHT   16.   (Part C., page C-3.)

conventional periodicity (atomic numbers 89-103);

(6) Measure of correlation

(a) between periodicity after Mendeleev and maximally-
hydrated weights (W_h, per Note 3, Part A) after Flint
(Exhibit C-14), and

(b) between these [(6)(a)] and prime numbers (Exhibit C-15).

Shakman, S.H., ATOMIC WEIGHT   17.   (Part C., page C-4.)

Exhibit C.

1. Flint, L.H.,J.WASH. ACAD.SCI.,*22*, No. 8, p. 99 (1932):

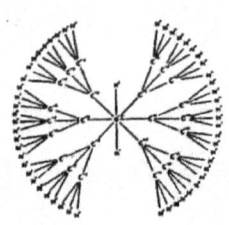

23 = postulated number of water molecules attached to fully hydrated ion of zero atomic number.

2. Van der Waals,J.N.& Platteeuw,ADV.IN CHEM. PHYSICS,*2*,p.11 (1959):

"Let us consider a clathrate crystal consisting of a cage-forming substance Q... The number of cavities of type "i" per molecule of Q is denoted by "v_i". ...for gas hydrates of Structure I /with 2 types of cavities/ v_1=1/23 and v_2=3/23...".

p.9: "If all cavities were filled, the composition would be 1 gas 5-3/4 H_2O...".

23 = number of water molecules per each 4 cavities in Structure I gas hydrate.

3. Berecz E.,& M.Achs-Balla,ACTA CHIM.(Budapest), *77*,p.277(1973):

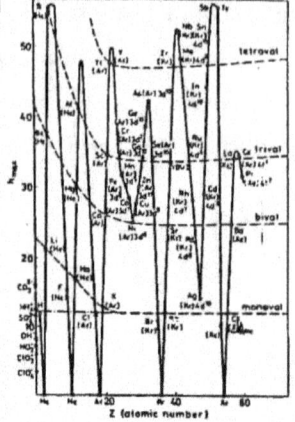

23 = "h_{max}" (maximum hydration number) indicated at atomic number zero by dotted line drawn through mono-valent "h_{max}" for Li, Na, and K ions.

4. Flint, L.H., HYDRATION AND BIOLOGY, p. 60 (1968):

Projected structural formula for carotin.

5. Miller, K.R., NATURE *300* (4 Nov. 1982), p. 53:

a. Individual three-dimensional model of individual membrane unit; b. cross-section of four unit cells (Rhodopseudomonas viridis).

6. Crewe, A.V., SCIENCE *221* (22 July 1983), p. 329:

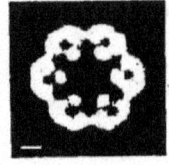

Top view micrograph of giant hemoglobin of an earthworm (Lumbricus terrestris)

Shakman, S.H., ATOMIC WEIGHT   18.   (Part C., page C-5.)

Exhibit C. (cont.)

7. after Flint, L.H., J.WASH.ACAD.SCI.,22. p.99 & 212-213 (1932):	8. Buhler, J.P., SCIENCE '85 (Nov.), p. 84-85:	9.Zhelonkin,A.V.,etal., SOVIET J.OF NUCLEAR PHYSICS,34(6)Dec.1981:	10.Einstein,A.,MEANING OF RELATIVITY;

Flint's Matrix –
   24 digits across:

H :23 22 21... 2 1 0

AN: 0 1 2...21 22(23)
   23 24 25...44 45(46)
   46 47 48...67 68(69)
   69 70 71...90 91 92

24=number of digits de-
   noting hydration
   number (H) and each
   of 4 overlapping
   periods.

6=number of circles
   fitting around
   circumference of
   circle, all of
   same radius.
12=number of three-
   dimensional balls
   that can simul-
   taneously touch a
   central one of
   the same radius.
24-25=number of four
   dimensional balls
   that can touch a
   central one.

24:(p.905)"24-plet of
   fermions" included
   in model of grand
   unification to study
   behavior of effect-
   ive charges; and

(p.908)in "5 mass
   matrices...Here, n 8
   n 24 n_0 16, count-
   ing separately the
   real and imaginary
   parts of F G and r."

Third Ed.,Appendix III:

24:In equasion for
   variation in
   planetary orbit:
$$+ \frac{24 \cdot 3a^2}{T^2 c^2 (1 - e^2)}$$

Fifth Ed.,Appendix II:

12=z_1 for pure gravi-
   tational field.
48=rejected z_1 for
   non-symmetric field
   (larger thus weaker
   than preferred
   alternative of 42).

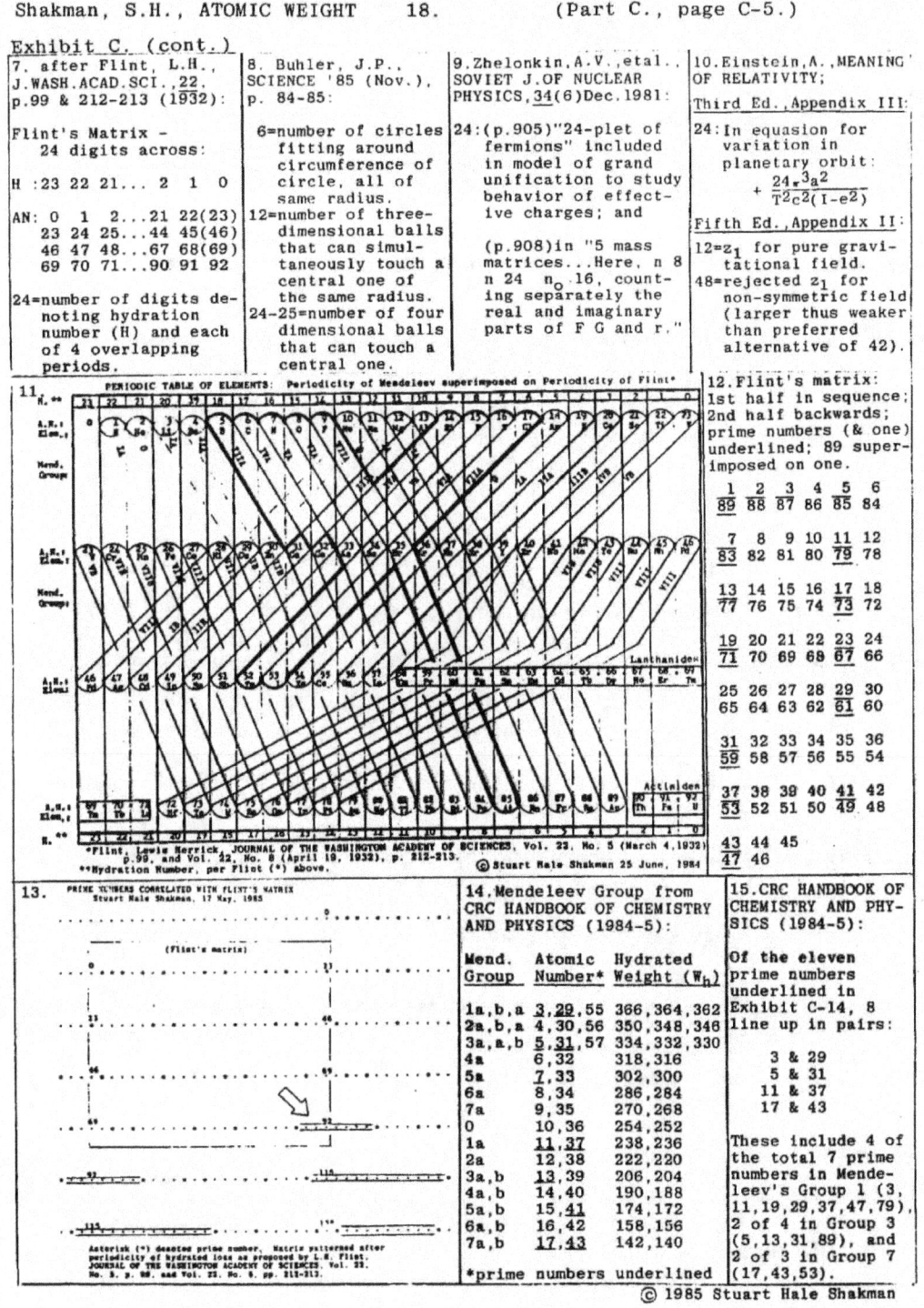

11. PERIODIC TABLE OF ELEMENTS: Periodicity of Mendeleev superimposed on Periodicity of Flint*

*Flint, Lewis Herrick, JOURNAL OF THE WASHINGTON ACADEMY OF SCIENCES, Vol. 22, No. 5 (March 4,1932) p.99, and Vol. 42, No. 8 (April 19, 1932), p. 212-213.
**Hydration Number, per Flint (*) above.
© Stuart Hale Shakman 25 June, 1984

12.Flint's matrix:
1st half in sequence;
2nd half backwards;
prime numbers (& one)
underlined; 89 super-
imposed on one.

1	2	3	4	5	6
<u>89</u>	88	87	86	<u>85</u>	84

7	8	9	10	11	12
<u>83</u>	82	81	80	<u>79</u>	78

13	14	15	16	17	18
<u>77</u>	76	75	74	<u>73</u>	72

19	20	21	22	23	24
<u>71</u>	70	69	68	<u>67</u>	66

25	26	27	28	29	30
65	64	63	62	<u>61</u>	60

31	32	33	34	35	36
<u>59</u>	58	57	56	55	54

37	38	39	40	41	42
<u>53</u>	52	51	50	<u>49</u>	48

43	44	45
<u>47</u>	46	

13. PRIME NUMBERS CORRELATED WITH FLINT'S MATRIX
Stuart Hale Shakman, 17 May, 1985

(Flint's matrix)

Asterisk (*) denotes prime number. Matrix patterned after periodicity of hydrated ions as proposed by L.H. Flint. JOURNAL OF THE WASHINGTON ACADEMY OF SCIENCES, Vol. 22. No. 5. p. 99, and Vol. 22. No. 6. pp. 212-213.

14.Mendeleev Group from CRC HANDBOOK OF CHEMISTRY AND PHYSICS (1984-5):

Mend. Group	Atomic Number*	Hydrated Weight (W_h)
1a,b,a	3,<u>29</u>,55	366,364,362
2a,b,a	4,30,56	350,348,346
3a,a,b	<u>5</u>,<u>31</u>,57	334,332,330
4a	6,32	318,316
5a	<u>7</u>,33	302,300
6a	8,34	286,284
7a	9,35	270,268
0	10,36	254,252
1a	<u>11</u>,<u>37</u>	238,236
2a	12,38	222,220
3a,b	<u>13</u>,39	206,204
4a,b	14,40	190,188
5a,b	15,<u>41</u>	174,172
6a,b	16,42	158,156
7a,b	<u>17</u>,<u>43</u>	142,140

*prime numbers underlined

15.CRC HANDBOOK OF CHEMISTRY AND PHY-SICS (1984-5):

Of the eleven prime numbers underlined in Exhibit C-14, 8 line up in pairs:

   3 & 29
   5 & 31
   11 & 37
   17 & 43

These include 4 of the total 7 prime numbers in Mende-leev's Group 1 (3, 11,19,29,37,47,79), 2 of 4 in Group 3 (5,13,31,89), and 2 of 3 in Group 7 (17,43,53).

© 1985 Stuart Hale Shakman

RADIO INTERVIEW TRANSCRIPT: "An Algebraic Approach to Unification
                              [aka 'The Secret Code of the Universe']

----

17 February 1987, 12 noon, at the 153rd Annual Meeting of the American Association for the Advancement of Science (AAAS, below), Hyatt Regency Hotel, Chicago.

Stuart Hale Shakman (SS, below), free-lance writer/investigator, interviewed by
Robynn Williams (RW, below) of Australian Broadcasting Corp. (radio).

Subject: "An Algebraic Approach to Unification"<A>, presented to the AAAS in
conjunction with printed 1987 AAAS Abstracts #110-113 [see <O>,<B>,<G>,<M>].

RW: What brings you to the AAAS? Keep playing if you like. What brings you to
the AAAS with your guitar?

SS: (playing guitar) This is one of my own [songs]. This is the FREEZING POINT
DEPRESSION BLUES. I'm here for the science convention. I've got four abstracts
that have been presented here at the convention, basically on the subject of
what happens when you dissolve substances in water.

RW: Why bring your guitar with you though?

SS: I was warned by the press people here that nobody might show up. So I
figured that, if nobody showed up, at least I'd get some practice in on my song.

RW: Presenting abstracts - is rather unusual. People usually present papers.
Why abstracts?

SS: The system here is actually pretty wonderful. It allows for an amateur
such as myself to submit [an abstract], as long as I get it into the proper
format, and have that [abstract] printed and become part of the official record.
  So the four [abstracts] that I did have printed here are part of the permanent
record, and they're also available for other researchers to use. And that's the
purpose of my printing them, in the hopes that other people will utilize them,
and hopefully build on this work, which I think is very constructive.

RW: Before I asked you that, I should have asked you before, your name, and
what you do, where you live.

SS: My name is Stuart Hale Shakman. I work part-time in a restuarant in San
Francisco -- it's Boz Scagg's restaurant as a matter of fact, the Blue Light
Cafe -- as a maitre d' and sometimes as a waiter, whatever it takes. And I
spend virtually all my free time on this research project that I have been
involved with for several years, 6 1/2 years.

RW: And the research project is about what?

SS: Initially I was looking into the origin of petroleum, and beyond that I
feel what I have discovered -- not discovered, nobody ever discovers something;
what you do is you stumble across something -- and I feel I have stumbled across
what I am calling "the secret code of the universe".

RW: "The secret code of the universe". Could you give us a brief outline?

SS: A brief outline is that this is a simple direct algebraic system ... that
seems to describe what happens when you dissolve substances in water,
mathematically, precisely. Now once we understand fully what happens when we
dissolve substances in water, we have a subject that's very important to

physics, to biology, and to chemistry. So we have something that unifies those
sciences. Thus far ... on conductivities and specific gravities ... I'm coming
very close, with ... calculations, to the actual values, just using atomic
numbers. It's a very direct means of calculation.

RW: Where did you pick up all this information, the skills, the understanding
of physics?

SS: I worked for the government for a number of years, the U.S. Government, and
in the government my work was involved with analyzing programs, analyzing
technical programs, analyzing their budgets and so on. At a young age, in my
20s, I was analyzing public works programs and agriculture programs for the U.S.
Agency for International Development.

RW: Yet I don't quite see how that can lead to the understanding of chemical
principles which directly relate to an understanding of the whole of the
universe.

SS: It relates to this because in any technical field, for the technical field
to be at all practical, a generalist has to take a look at it, analyze it and
translate that out to your public. I had the experience of working with that in
the U.S. Government, so I wasn't afraid to tackle a project like this, even
though I am an independent; I am not affiliated with any university. I'm an
analyst. I analyze numbers. I "crunch" numbers.

RW: Sure, there's one point, however: If it is of such moment that it gives
you an answer to the universe -- and I did notice that you had various names
like Einstein, Newton and so forth -- if you give this paper, or the abstracts,
isn't that "it" for science? I mean if the solutions are there ...?

SS: It's just a beginning. ... It's a framework to begin to understand ...
There are different options. The system tells you how many water molecules
attach to different ions. But in some cases ... they don't get any water
molecules. In some of the calculations, I have [assumed that] ... half of them
get all the water molecules they're supposed to get and half don't get any at
all. ...If I make that assumption and then I do my calculations, I come out
right on the money. So maybe that says that this is what is happening.

RW: What if nobody comes to hear your paper. Will you be terribly disappointed?

SS: I would feel negligent if I did not at least give them an opportunity.

RW: No, I'm just thinking that I'm the only person who has actually come to
your display so far here at the AAAS ...?

SS: It really doesn't matter. I've come across something that's really
beautiful, and I've decided that I'm going to use all my energies to share this.

RW: Could you give us a little more music before we go?

SS: Sure (playing more FREEZING POINT DEPRESSION BLUES). ... This system has
been around for several billion years. ... We're just finding out about it --
whether we do it today or tomorrow. We're just agents of acceleration; we can't
change what is (finishing up FREEZING POINT DEPRESSION BLUES).

RW: That's delightful. Thank you.

SS: Thank you.

7777 Fay Ave., #K-292
La Jolla, CA 92037
(619) 454-3132
13 December 1989

Dr. Robert White, Director
THE EXPLORATORIUM
Lyon Street and Marina Blvd.
San Francisco, Ca.94123

Dear Dr. White::
    Thank you for taking the time to meet with me on Wednesday, 9 August 1989. Your
thoughtful questions have been very helpful, as I hope will be evident in this
attempt to address them.

    Regarding the physical model on which is based Flint's methodology, Flint cited
the so-called Graham law of effusion/diffusion (for gases), which he extended to
mobilities of solute ions (as indicated by their electrical conductivities) in accord
with van't Hoff's analogy between gases and solutes. (This is the same van't Hoff
who after death [1911] was succeeded by A. Einstein at the Royal Prussian Academy of
Science in Berlin [1914]).
    A direct relationship between mobility and conductivity has long been
recognized, e.g., as in F. Kohlrausch's 1876 work [GOTTINGEN NACHRICHTEN, 1876, 213,
per H.M.Goodwin, SCIENTIFIC MEMOIRS...] and in Einstein's "Elementary Theory of the
Brownian Motion" of 1908 [per BROWNIAN MOTION, Dover 1956, p. 84; this work also
involved electrical conductivities of hydrogen and potassium ions, as did Flint's
initial work on hydration in 1932.] Flint reported that he was not aware of any
attempt, prior to his own, to extend the Graham principle to solutes, nor am I.
    Of the relationship of Graham's law to the kinetic theory, Linus Pauling [
CHEMICAL BOND, 1967, p. 174] has noted:
"...the kinetic theory requires that the rate of effusion of a gas through a small
hole be inversely proportional to the square root of its molecular weight. This law
was discovered experimentally before the development of the kinetic theory ...".
    Thus have relative mobilities of gases been viewed as indicators of relative
weight, and so might mobilities (or electrical conductivities) of solute ions, in
accord with the kinetic theory, be viewed as (quite precise) "scales" for measuring
the relative weights of the (possibly "hydrated") ions in solution. This is what
Flint did. He used conductivities to "weigh" ions.

    Against the relative conductivities of Li+, Na+ and K+ ions, Flint balanced
their presumed hydrated weights, comprised of atomic weights plus the weights of
however many water units would have to be added to each ion, so that the inverse-
square-roots of their respective new total weights were in the same proportions as
were their conductivities. Flint's (proposed) description is a simple and direct
extension of this initial calculation.
    Specifically he calculated that 4, 12, and 20 water units (with weight values of
18 each) would need to be added to weight values of 39 for K+, 23 for Na+ and 7 for
Li+, respectively, so that the relative inverse-square-roots of resulting total (
hydrated) weights of the 3 ions would match their relative conductivities (65.3,
44.4, and 35.5, after Nernst per W.M.Bayliss [PRINCIPLES OF GENERAL PHYSIOLOGY 1915,
p. 177]).
    A review of the body of information presented by Bayliss may give us a clearer
picture of why Flint said his "discovery took place without the involvement of any
extraordinary mental prowess" [BEHAVIOR PATTERNS, 1964, p. 17]:
 - Bayliss [1915], p. 177: refers to Nernst conductivities of 65.3 for K+, 44.4 for
Na+ and 35.5 for Li+;

- Bayliss [1915], p. 176: "… we are struck by the fact that lithium, with an atomic weight of 7, moves at a much slower rate than potassium, with an atomic weight of 39.

The explanation is probably that the lithium ion carries with it a larger number of water molecules than the potassium ion …"

-Bayliss [1915], p. 178: "The chief work on this question has been done by Bousfield (1905, 1906, 19112)" who proposed hydration numbers for K, Na and Li of 4, 8 and 16, "As the author says, 'an attractive looking series.'" [Bousfield's work is cited in relatively contemporary literature in the context of, and preceding, that of Born: Rosseinsky, D.R., CHEM. REV. 65 (1965), 467].

Let's walk through Flint's claimed "discovery", treating electrical conductivities literally as a type of electrical "scale". If, in accord with Graham (and the kinetic theory), velocity (or conductivity) varies with the inverse-square-root of weight, then weight would vary as the inverse square of velocity (or conductivity). Thus the relative total weights of the hydrated ions could be characterized as 1/65.3-squared) for K+; to 1/ (44.4-squared) for Na+' to 1/(35.5-squared) for Li+; or .000234 : .000507 : .000793.

If we assume that the K+ ion has 4 associated water units as proposed by Bousfield, its hydrated weight might be calculated as the sum of its atomic weight (39) plus the weight of 4 water units (18 each) for a total hydrated weight of 111. The the relative calculated weight of the hydrated Na+ ion would be 239 [111(.00507/.00235)] and that of the hydrated Li+ ion would be 375 [111(.00793/.00235)], weights that could most readily be approximated by assuming Na+, with an atomic weight of 23, is associated with 12 water units; and li+, with atomi weight of 7, is associated with 20 water units. (While Flint did not mention Bousfield, it is tempting to speculate that he first calculated theoretical relative conductivities using Bousfields's proposed hydration numbers, and then adjusted these numbers to better correspond to the conductivities. If so, then Flint's "discovery" was delivered through a fairly narrow window in the history of opportunity, in that reference to Bousfield's proposed hydration numbers was omitted from a subsequent [1924] Bayliss edition.)

Flint noted that the sums of atomic numbers and his calculated hydration numbers were in each case 23. He then refined and extended this pattern, in accord with updated and more comprehensive SMITHSONIAN PHYSICAL TABLES data, into a proposed system involving 4 periods (N= 1-4) of 23 elements each, comprising the total of 92 naturally-occurring elements, wherein the (maximum) hydration number (H) might be calculated by the equation: $H = 23N - (Z+C)$; Z=atomic number, C=valence, H= 23 to 0, N= 1 to 4.

In his refined and extended system, Flint proposed a unitary change in atomic number (Z) for each unit charge (C), whereby he was proposing a change of weight with ionization. Of this he stated: "It was a drastic revolutionary and seemingly incredible step, -- but it permitted an integration of observational data with a satisfying convincing nicety" [BEHAVIOR PATTERNS OF HYDRATION, 1964, p. 19]. Flint accordingly characterized water of hydration not as presumed neutral H2O molecules, but rather as (negatively charged) H2O- ions, with atomic "weight" of 18 on a scale of 16 for the theoretical, neutral, oxygen atom.

The hypothesis that an entity may weigh other than the sum of weights of constituent atoms recalls the suggestion made by Marignac in 1860 in comments supporting Prout's hypothesis, that a grouping of primordial atoms in the form of a (more complex) chemical atom might weigh other than the sum of weights of constituent atoms; similarly, in 1921 F.W. Aston proposed the concept of a "packing" effect which might allow for an atomic nucleus to weigh less than the sum of weights of constituent "charges". [Prout, W. (1815-6), J. Stas & C. Marignac (1860), PROUT'S HYPOTHESIS, p. 58, 22.]

We might also speculate that Flint, who was familiar with the work of S. Arrhenius, may have borrowed from Arrhenius' (out of context) statements that inherent in Mendeleev's scheme, "atomic weight ought to increase with the positive valence" [of consecutive elements] and that Thompson's scheme would require the

supposition that a quantity of electricity "makes the difference between two
consecutive elements". [THEORIES OF CHEMISTRY (1907),pp.100-1])

     Regarding your question concerning the significance of electrons associated with
any given atom or ion as may be related to hydration, the methodology proposed by
Flint involves only atomic number and valence (a single negative charge subtracts one
from the "neutral" atomic number, and a positive charge adds one). This does not
mean that other information concerning the electrons is without value, but rather
that even before such information is factored in we have at our disposal a computer-
friendly tool for deriving first approximations of various measurements concerning
the still-invisible and largely unknown world of solute behavior.
     That such an approach to understanding solute behavior is available is not
generally known. For example, in the case of mobility (or conductivity), M.
Berkowitz and W. Wan stated in 1987: "The mechanism of ionic transport is a
classical problem in physical chemistry. ... To describe the limiting ionic mobility
on a molecular level is a very challenging task. It is obvious that first theories
will have to include simplifying assumptions..." [J. CHEM. PHYS., 86 (1 January 1987)
376-382].

     The periodicity of hydration may at first glance seem an excessively narrow
field; however,
     (1) solutes (and hydration) are intimately involved in metabolic processes, so that
an underlying periodicity, if such exists, would be of great power and importance;
     (2) Flint's methodology alone allows for direct approximations of hydrational
behavior; including both diameters and weights for the same set of hydrated ions ["A
Lock for Flint: Diameter & Conductivity" PROCEEDINGS AAAS PACIFIC DIVISION 7 (1988),
42]*, specific gravities [after Flint 1964**, as per 1987 AAAS Abstract #113]*;
experimental osmolalities [Flint 1964, 1968]**; published (CRC) maximum osmolalities,
solute diffusion and apparent equilibrium distribution coefficients [unpublished data
]; all using only atomic numbers, valences and algebraic formulations;
     (3) patterns encompassing lanthanides, actinides and nonmetals, resulting from an
overlay of Mendeleev's periodicity on Flint's ["Mendeleev Groups overlaid on Flint
Periods.", 1987 AAAS Pac. Div. Proceedings, Vol. 6, Part 1, p. 39]* argue for primacy
of Flint's periodicity;
     (4) Flint [1973]** provided some speculative work on extending his methodology to
the solid state; and
     (5) this methodology also allows for direct approximations of gaseous
interdiffusion, which approximations compare favorably with Hirshfelder, Spotts and
Bird's (indirect) calculations of interdiffusion from viscosity ["Weight of H2 Gas",
1986 AAAS Abstract #212]*; cathode ray absorption-to-density ratios for H2 gas
relative to other gases and solids [1987 AAAS Abstract #110]*; offers an approach to
understanding the behavior of ionized gases [Flint 1964]**; and offers an explanation
as to why H2 and He gases have essentially the same lifting power [CRC HANDBOOK, 1985
-6, p. B-20; discussed in 1987 AAAS Abstract #110]* (such would be the case if H2 and
He gases weighed the same).
     (The equivalence of lifting power for H2 and He gases highlights a most
fundamental (and bound-to-be controversial) aspect of Flint's work - - his assertion
of the primacy of atomic number over atomic weight as a true measure of relative
weight of (theoretically "neutral") atoms. It may well be emphasized that this is
not a prior assumption on which Flint's work was based, but rather one which
logically followed from his calculation of (unitary) hydration numbers; and is
consistent with the position that "the atomic number is a more fundamentally
important property of an element than its atomic weight [J.C.Speakman, MODERN ATOMIC
THEORY, 1938, p. 83].)
     In summary it may be offered that the periodicity of hydration, as described by
Flint, encompasses that of Mendeleev (so nothing is lost), progresses in exploring
some of the unsolved mysteries inherent in Mendeleev's periodicity, and goes on to
tackle problems beyond range of Mendeleev's periodicity.

It was illuminating to learn that the simplicity and mathematical structure of Flint's methodology as might render it vulnerable to characterization as a form of "numerology". Ironically, if the proposed description of Flint is found indisputably to be descriptive of reality, undoubtedly there will be persons who will strive to extend its significance into the occult (Flint would not have approved).

Nonetheless, investigators of natural phenomena have long sought to understand their mathematical underpinnings, which quests were not necessarily derived from belief in the occult. Thus had Newton consolidated the mathematical foundations of science up to his time within his "Mathematical Principles". Still, Newton looked forward to the day when a more complete mathematical description would be known: "I am induced by many reasons to suspect that they [the phenomena of nature] may all depend upon certain forces by which the particles of bodies by some causes hitherto unknown are either mutually impelled towards one another, and cohere in regular figures, or are repelled and recede from one another." [Newton, I., MATHEMATICAL PRINCIPLES (1687), Preface to the First Edition].

Moreover, in A. Einstein's last published edition of RELATIVITY, he emphasized what he saw was a need "to find a purely algebraic theory for the description of reality." [RELATIVITY, Princeton, 1956].

As Max Born subsequently noted, modern researchers responded to Einstein's call for unification, "though not in his conviction that the unified laws ought to be of the classical, deterministic character" [EINSTEIN'S THEORY OF RELATIVITY (1962) p. 372].

Nigel Calder's perspective on Einstein borders on the abusive, going so far as to suggest that poor Albert had degenerated into a religious fool: "He [Einstein] developed one appalling blind spot ... His vehement rejection of [the quantum theory] was in part at least a consequence of his pantheistic belief in a perfect universe ... Einstein did not seem to realize that the perfection he craved for would be sterile ... He spent his last thirty years trying to unify gravity and electricity - but electricity could be attacked properly only in terms of the quantum theory. ... The young Einstein might have managed something, but not Einstein past his peak ... As individuals, scientists can be as pigheaded about their ideas as anyone else ...", etc. [EINSTEIN'S UNIVERSE, The Viking Press: New York, 1979, pp. 140-1].

Friends of Albert Einstein may find some measure of comfort in the prospect that Einstein will have the prophetic last word against his detractors, should Flint's proposed (algebraic) description hold up. Further, are not both (a) Flint's extension of the Graham principle (or law of kinetic energy), so as to describe the relation between relative respective weights and electrical conductivities of ions in solution; and (b) his hypothesis of weight change with ionization; intimately related to the question of "unification" of gravity and electricity?

Thank you for your expression of willingness to consider a specific proposal if submitted. While the system underlying the methodology described by Flint is actually quite simple (far less complex in any case than contemporary methodologies which moreover don't work as well), explanations as to why it merits consideration as at least a model of scientific reality can get complicated.

The August 1989 Journal of Chemical Education's editorial quotes Percival Lowell, predictor of Pluto's existence, as having asserted: "to set forth science in a popular, that is generally understandable, form is as obligatory as to present it in a more technical manner." In this spirit would I welcome the challenge of bringing the methodology of Flint within the grasp of the EXPLORATORIUM'S audience, particularly as this encompasses secondary and college levels. If (when) this methodology can be presented directly as a model, without getting involved in a bunch of "red herrings" (a term with which former EXPLORATORIUM Professor Paul Hewitt deflected questions that would otherwise divert him from the main point he was discussing), its most basic assumptions can be introduced to anyone who can count to 92; its periodicity to anyone who can multiply 23 times 4; and its full complexity to anyone who can comprehend inverse-square-roots and cube-roots. An exhaustive assessment of Flint's methodology has convinced me that this will eventually occur, and exposure through the EXPLORATORIUM has been sought in the hopes of accelerating this eventuality. Moreover an early association of the Exploratorium with this methodology, if it indeed turns out to be all that is herein claimed, could also serve to enhance the role of the Exploratorium in the future of the scientific enterprise.

Enclosed is a copy of Alfred Jensen's "Solar Energy Optics" [1975] as exhibited in the March/April 1988 issue of THE SCIENCES; which painting bears an uncanny resemblance to a matrix of elements based on Flint's methodology, as exhibited in materials previously sent to you; which painting and matrix conveniently have in common a science-unification theme. It is doubtful Jensen was familiar with Flint (one would think in this case that there would be more identifiable detail and an appropriate citation), and in any case Flint did not display his periodicity thusly. Nor was I familiar with Jensen prior to seeing the 1988 THE SCIENCES article. Perhaps the Jensen painting might be obtained on loan, and along with the matrix/ quadruple-helix after Flint [as indicated to you the matrix, when wrapped into a cylinder so that the atomic numbers run continuously, becomes a helix]* and a simplified explanation of Flint's methodology, be incorporated into an "ARTISTS IN RESIDENCE" exhibit. Other Jensen paintings and additional graphic aspects of Flint's work, plus Buckminster Fuller's "closest-packed-spheres", might also be incorporated into such an exhibit.

Alternatively the Flint methodology might be given a more complete showing in the form of a video, which video project might also incorporate the Exploratorium's "Periodic Table" and other appropriate exhibits. As currently envisaged, this video would be entitled "THE SECRET CODE OF THE UNIVERSE", would be built around explanatory interaction with a child or group of children, and might also involve an inter-active component entitled "The Ultimate Computer Game". A draft outline for such a video project has been prepared and can be made available if appropriate staff at the Exploratorium would be interested in pursuing this further.

As of December 18 I am returning to Southern California and can be contacted at the La Jolla address shown at the top of this letter. Correspondence with your successor and/or appropriate staff concerning a potential EXPLORATORIUM-sponsored exhibit or video would be welcome. Depending on the availability of necessary funding I would be interested in visiting San Francisco for whatever period of time is required to work with your staff in fine-tuning and implementing such a project. Also most welcome would be any comments or suggestions you might offer concerning the adequacy of the above presentation in addressing the points you have raised, along with your forwarding address.

Thank you again for your kind and courteous attention, notwithstanding the admittedly unorthodox and controversial nature of the subject matter brought to you. Best wishes to you in your future endeavors and a happy holiday season to you and your wife.

With kind regards,

Sincerely yours,   Stuart Hale Shakman

* Included in 3 May 1989 letter to you or hand-delivered on 9 August 1989.
** Flint, Lewis H., 1932: JOURNAL OF THE WASHINGTON ACADEMY OF SCIENCES 22, 97-119, 211-217, 234; 1964: BEHAVIOR PATTERNS OF HYDRATION; 1968: HYDRATION AND BIOLOGY; 1973: DISSENTING APE

SUBMISSIONS TO NATURE & ASSOCIATED LETTERS (August 1992-Feb. 1994)

The following two articles ("Conductance, Hydration and Periodicity" and "Not Just A Flake Model") are admittedly largely duplicative, with correspondence in the middle trying to explain the reason for the second (revised) article.

For convenience, the parts of the respective articles that are identical or nearly so are highlighted in both articles, so that the items/ narrative not duplicative can be readily discerned.

The following article was submitted, Aug 15, 1992; assigned #SO8279

PROPOSED ARTICLE

Conductance, Hydration and Periodicity

Stuart Hale Shakman

P. O. Box 382, Santa Monica, CA, U. S. A. 90406-0382

Albert Einstein, 1944: "Even the great initial success of the quantum theory does not make me believe in the fundamental dice-game, although I am well aware that our younger colleagues interpret this as a consequence of senility. No doubt the day will come when we will see whose instinctive attitude was the correct one"[1].

Flint's classically-based methodology[2,3], correlated herein with various solute phenomena and Mendeleev's periodicity, counters quantum approaches to conductance[4,5] and supports Einstein's vision of algebraic unification[6a].

IN that aqueous solution dynamics underly many important physical, chemical and biological processes[4], implications of the late Lewis H. Flint's use of Graham's law to interpret conductance data as a measure of relative solute ionic weight[2] may interest a wide range of investigators. Flint's calculations appeared to disclose simple algebraic patterns of hydration which he described as based on first principles of atomic number (Z) and valence (C)[2,3a]. So based and with so-called "heuristic value", Flint's methodology meets some criteria for consideration, if not "as a starting point for the calculation of everything", at least as an intermediate model of reality[7].

This paper discusses circumstances culminating in the development of Flint's methodology and illustrates its potential utility. For 100% of the 22 CRC-listed[8] mono-atomic positively-charged elemental ions with $Z = 1$-19 or 55-80, plus OH-, observed conductivities are within 10% of calculated values, assuming either maximum or zero hydration as per Flint (Table 1); calculations for both H+ and OH- ions involve the assumption of zero hydration, indicating that this and not "quantum effects"[4] may be primarily responsible for their high conductivities.

Also discussed, and based on this same algebraic methodology, are agreement with alternately-determined relative hydration values (e.g. as in Table 2), approximations of both linear size and conductivities for two sets of ions (Table 3), correlations with relative energies of solution (Table 4), patterns corresponding to Mendeleev's periodicity (Table 5), and a proposed explanation for the six-cornered snowflake.

Background

While he is best known for his work on the periodicity of the elements, Dimitri Mendeleev was also, by 1887, an early and strong supporter of the hydrate theory of solutions[9]. Bredig's subsequent use of the hydrate concept in 1894 to explain why the chlorine ion is not more mobile than the heavier iodine ion, attributing this to a larger hydration shell around the Cl- ion that migrates with it[10], has remained essentially unchanged up to the current time[11].

In 1899 Abegg and Bodländer sought to explain why mobility of the K+ ion is greater than that of the smaller and lighter Na+ ion, and the latter greater than that of the even smaller and lighter Li+ ion, with the suggestion that the hydration potential of these ions may vary inversely with their anhydrous weights[12]. This would allow for the hydrated Li+ ion to be the heaviest and least mobile of the three ions, and the much-less-hydrated K+ ion to be the lightest and most mobile. In accord with this same premise, they offered that the large conductivities of H+ and OH- might be attributed to an absence of hydration. Over the next few decades, other investigators similarly invoked the concept of hydration in discussions of conductivities[13-16]; however, it was not until 1932, when Flint called on Graham's law[17] to deliniate the relationship between hydration and conductivity, that a mathematical framework for Abegg and Bodländer's suggestions was advanced, exposing an apparent algebraic relationship between solute ions and their water solvent.

In accord with Graham's law of diffusion (or effusion, of gases), the mobility of a gas varies inversely with the square root of its (density or) molecular weight[17], a relationship required by the more recent law of kinetic energy[18] and one which may be derived from the right side of Newton's equation characterizing resistance as varying "directly as the squares of velocity and ... as the quantities of matter"[19a]. (Graham himself credited Professor John Robison with having deduced this "pneumatic law" directly from the pre-Newtonian theorem of Torricelli on the velocity of efflux of fluids[17].) As solute behavior had been shown by van't Hoff to be analogous to gaseous[20], and conductivity was recognized as an index of mobility[21,22], Flint was able to view conductivity values as a measure (inverse-square-root) of relative hydrated weights of solute ions. In essence, Flint's work rendered the Wheatstone bridge, which measures conductivities, a "scale" for "weighing" solute ions.

(Van't Hoff's successor at the Royal Prussian Academy in 1914, Albert Einstein[23], had as early as 1907 used the relationship embodied in Graham's law in conjunction with solutions, characterizing velocity as varying with the inverse-square-root of mass for particles in colloidal platinum solutions[24]. But Einstein apparently did not attempt to extend this characterization to considerations of conductivity, although he did refer to conductivities of H+ and K+ ions in a 1908 discussion of "displacement"[22]. Nor did Walter Nernst[15], who was eulogized in 1933 by Einstein as "having ascended from Arrhenius, Ostwald and Van't Hoff, as the last of a dynasty which based their investigations on thermodynamics, osmotic pressure and ionic theory"; Nernst was specifically remembered by Einstein for his "witty" use of the Wheatstone Bridge[25].)

In 1932[2] Flint cited Nernst conductivity values for K+, Na+ and Li+ of 65.3, 44.4, and 35.5, resp., as listed W. Bayliss's 1915 physiology textbook[14]. (On an adjoining page of this text were listed W. Bousfield's proposed hydration numbers 4, 8 and 16 for these same ions, a self-proclaimed "attractive looking series"[26].) Inspired by Abegg and Bodländer[3b], Flint sought to determine hydration numbers that would allow for an optimum fit with the Nernst data, via Graham's law. He found that presumed hydration numbers of 4, 12, and 20, each multiplied by 18 (for the weight of each water unit) and added to conventionally- appraised atomic weight values of (approx.) 39 for potassium (K), 23 for sodium (Na) and 7 for lithium (Li), resp., yielded totals (111, 239 and 367) whose relative inverse-square-roots (.0949, .0647 and .0522) correlated remarkably well with the Nernst data.

As his initially-hypothesized hydration numbers (4, 12, and 20) were integers, each of which when added to its associated atomic number (19, 11 and 3 for K, Na and Li, resp.) equalled 23, Flint was induced to attempt to use integral values based on atomic numbers in his conductivity calculations[3c]. He then found that adjusted atomic-number-equivalent values (Z') which incorporated a shift equal to one unit Z per unit valence (C) "permitted an integration of observational data with a satisfying convincing nicety"[3d] (equation 1):

$$Z' = Z + C, \qquad\qquad (1)$$

where Z=atomic number and C=valence. (Flint's use of values based on atomic numbers is consistent with contemporary recognition of the primacy of the atomic number[27] as established by Moseley[28]; contemporary science also embraces the concept of a shift in atomic-number-equivalent values with ionization in the case of an increase in the atomic number of some radioactive elements resulting from the loss of a nuclear electron[29].)

Noting that the number 23 equals one-fourth of the number (92) of naturally-occurring elements, Flint projected four equal periods, in each of which the maximum hydration number (Hmax) decreases from 23 to zero with increasing atomic number[2,3e] (equation 2):

$$\text{Hmax} = 23n - Z', \text{ when Hmax} = 23 \text{ to } 0, \text{ and } n = 1 \text{ to } 4 \qquad (2)$$

(for $Z' = $ 0 to 23, n=1;

for $Z' = $ 23 to 46, n=2;

for $Z' = $ 46 to 69, n=3;

for $Z' = $ 69 to 92, n=4).

In his calculations Flint characterized water units involved in hydration as negatively-charged ions $(H2O-)$[3e], allowing for calculation of the relative hydrated Z' value $(Z'h)$ for a given solute ion as the sum of Z' for the ion plus a value of 9 for each associated H2O- unit (equation 3)[3a]:

$$Z'h = Z' + 9H, \qquad (3)$$

where H is the actual number of associated H2O- units.

Approximations of Ionic Data

Table 1 displays observed conductivities for the 41 monoatomic elemental ions through Z=80, plus OH-, listed in CRC[8] and error values (E) for approximations derived as the relative inverse-square-root of $Z'h$, under one of four hydrational assumptions - maximum, zero, half-maximum and one-fourth-maximum hydration levels. (Throughout this paper, "maximum" or "full" hydration levels are maximum levels as proposed by Flint.) Maximum and zero hydration were the primary assumptions employed by Flint, but he also introduced the concept of shared water ions within discussions of hydrational bonding and utilized fractional-maximum-hydration levels in associated calculations[3f].

The specific impetus for the use herein of a speculative half-maximum-hydration assumption is the observation that, as measured against presumed maximum hydration for the base ion Na+, calculations of both conductivity and diameter for the Cl- ion are optimum using an hydration number of 3.5 (as in Table 3), a value first proposed by Bousfield in 1906[30]. It was noted that this value is equal to half the maximum hydration number as per Flint, and further that a half-maximum-hydration assumption allows for agreeable conductivity calculations for F- and for the 6 divalent "transition-series" ions Mn++, Fe++, Co++, Ni++, Cu++ and Zn++. Thus conductivities for all but 7 of the 42 ions in Table 1 are within 10% of calculated values, assuming either maximum, zero or half-maximum hydration levels. A single fourth (speculative) assumption of a one-fourth-maximum hydration level allows for approximations of conductivities for 5 of these 7 with error values less than 10%.

(It may be noted that each of the four hydrational assumptions as utilized in conductivity calculations in Table 1 is associated with a distinct grouping of atomic-number-equivalent values [Z'], which groupings tend to correspond with Flint's periodicity: The assumption of maximum hydration is associated with all six CRC-listed[8] univalent positive ions with Z'=4 to 20, all within Flint's first hydrational period; half-maximum with all six divalent "transition-series" ions with Z'=27 to 32, all within Flint's second hydrational period; one-fourth-maximum with all three CRC-listed[8] trivalent "transition-series" ions with Z'=24 to 29, all within Flint's second hydrational period; and zero hydration with all of 16 CRC-listed[8] ions with Z'=52 through 82, values falling within Flint's third and fourth hydrational periods.)

Calculations of relative mobilities of ions, aside from playing an indispensable role in Flint's initial work on hydration, were subsequently used by Flint in discussions of the concept of "Ph" and development of a "tentative" acidity-alkalinity index[31,3g].

Although efforts to determine hydration numbers have yielded a wide range of values, the ratio of values for Na+ v.s. Li+ per Flint (11/19 = .58) correlates reasonably well with the average (.61) for 14 sets of values by 9 investigators as listed by Amiss[32]. Even more striking is how Gluekauf's relative hydration values[33] for the series H+, Li+, Na+, K+, Rb+ and Cs+, adjusted to a base value of 11 for Na+ as in Table 2, correlate precisely with Flint's assumed maximum (H+, Li+, Na+ and K+) or zero (Rb+ and Cs+) hydration levels; except for H+, all adjusted values in Table 2 are identical with hydration values used in Table 1.

Seeking to validate his work through the approximation of specific gravities, Flint proposed that the relative volume of a solute ion (VZ) may be approximated as the quotient of (a) Z'h and (b) $1 + Z'/Z'h$ (equation 4)[3h]:

$$VZ = Z'h/(1 + Z'/Z'h) \qquad (4)$$

Flint used this formulation in deriving theoretical values for comparison with specific gravity data[3h], results of original osmosis experiments[3i], gas solubility data[34a], and densities of elemental solids[35a], but is not known to have used it in conjunction with linear ionic size.

Table 3a lists observed relative diameter values for a set of 8 solute ions of physiological interest[36], observed conductivities, and approximation-error values for both sets of data. Diameters are approximated as the cube-root of VZ, relative to a base value of 1.00 for Na+; conductivities are approximated as the inverse-square-root of Z'h, relative to 50.08 for Na+ as in Table 1.

As shown in Table 3a, observed diameters for 7 of the 8 listed ions (all except HPO4--), and observed conductivities for all 8 ions, are within 11% of calculated values, assuming a "maximum" hydration level in all cases except (a) a half-maximum-hydration assumption used in calculations of both diameter and conductivity of Cl-, and (b) a zero hydration assumption used in calculation of conductivity of SO4--. In the case of HPO4--, conductivity has been approximated using the assumption of maximum hydration, but diameter may be approximated (as relative cube-root of VZ) only if ionic size is assumed to be even larger -- approximately equal to that of two fully hydrated ions. (Such an hypothesized double ion might then be projected as splitting under electrical stress into separate, "fully" hydrated ions.)

Table 3b lists observed crystal ionic radii[37] and approximation error values for the series Li+, Na+, K+, Rb+ and Cs+. Radii are approximated as the relative cube root of theoretical volume, as were diameters in Table 3a, except that for all ions in Table 3b approximations of radii involve the assumption of zero hydration. While the order of agreement for approximations of linear size is less satisfactory in Table 3b than in Table 3a, it may be noted that the trend in error values in Table 3b for Li+, Na+ and K+ radii is consistent with Flint's discussion of how the attractive force mediating hydration may also act as a relative cementing force in solids[35a]; this might allow for the Li+ ion, with a greater hydrational/attractive potential than Na+, to be more compressed in crystal form, and the K+ ion, with a lesser potential than Na+, to be less compressed. (Conductivity data and approximation error values from Table 1 are duplicated in Table 3b for convenient reference.)

Hydrational potentiality per Flint is correlated with components of energies of solution[38] in Table 4. Relative square-roots of Z'h values associated with maximum hydration for Li+ and Na+ against half-maximum hydration for F- and Cl- (assumptions used in Table 1) correlate reasonably well with "total solvent energy" values ($\wedge Esol$); whereas relative square-roots of Z'h values associated with "maximum" hydration for all 4 ions yield comparably-satisfactory approximations of the "bulk solvent" portion (ESX2) but not first shell portion (ESX1) of total solute-solvent energy (ESX; ESX = ESX1 + ESX2). For the first shell (ESX1) a shift in assumptions for the Cl- ion, from a maximum to a half-maximum hydration level, yields an improved correlation for Cl- relative to calculations based on full hydration for the other three ions.

Periodicity

In attempting to correlate Flint's periodicity with that of Mendeleev, it was noted that Z values associated with three (Mendeleev) Group 1 elements (Z = 3, 29 & 55) are associated with consecutive Z'h values (Z'h = 183, 182 & 181, resp.) Consequently, Z'h values for all values of Z from one through 92 have been calculated and exhibited in order of ascending Z'h values, as in Table 5. (Z'h values are calculated as per equation 3, assuming full hydration as per equation 2 and C=0 in equation 1.) These Z'h values fall into groupings of up to four consecutive values (e.g., 148, 149, 150, 151), each value 8 digits distant from the same position in the next grouping (156, 157, 158, 159).

Data in Table 5 is arranged in five columns with a difference between Z'h values horizontally of 64, which arrangement allows all atomic numbers from Z=2 through Z=57 that fall within Mendeleev's Groups 0 through VII, except Z=24 & 25, to fall into corresponding horizontal groupings. The Z value for hydrogen (Z=1) is found with Z values for elements in Mendeleev's Group VII, where hydrogen is sometimes categorized due to some recognized chemical similarities[39]. (A unique sort of balance is noted within the horizontal grouping encompassing Z = 2, 10, 18, 36 and 54, atomic numbers associated with Mendeleev Group 0: The Z'h value for Z=18 is 63, and all other Z'h values in this horizontal grouping are approximate multiples of this value. Also noted among horizontal groupings in Table 5 is that, for Z=2 through 57, none of six prime-numbered Z values in Mendeleev's Group I is paired with a Z'h value which is also a prime, whereas two of three prime Z values within each of Mendeleev Groups III, V & VII are paired with Z'h values which are also prime numbers.)

Flint's periodicity also dovetails nicely, geometrically, with that of R. B. Fuller: As per equation 2, Flint's four (overlapping) hydrational periods of 24 digits each (including "0" and duplicates of "23", "46", and "69"), terminating in the number 92, bears an uncanny resemblance to R. B. Fuller's "four of the 24-ness of the duo-tet cube" (and "disappearing octa set" of one "expendable exterior octa" and "three expendable interior octa")[40]. A prospective "disappearing octa set" is virtually gone when Flint's periodicity is characterized continuously and three-dimensionally as a quadruple helix.

Conclusions and Speculations

Beyond arguing for consideration of Flint's methodology as may relate to a general understanding of solute phenomena, this paper specifically supports (1) Bousfield's hypothesis of an hydration number of 3.5 for Cl- (incident to measurement of conductivity and diameter as in Table 2), offering that this may represent a level equal to one-half the maximum hydration level per Flint and further that conductivities of several other ions may similarly indicate hydration levels which may be derived as distinct fractions of respective maximum hydration levels; (2) the hypothesis that the sulfate ion, fully hydrated when measured for diameter, may shed its water complement under electrical stress incident to measurement of conductivity; and (3) the work of Abegg and Bodländer[12], and others[15,16], suggesting that the large conductivities of the OH- and H+ ions in solution are due to a relative absence of hydration, although these ions may otherwise be strongly hydrated as has been inferred from studies of specific gravities of acidic aqueous solutions[35b] and relatively large heats of hydration[5].

The ease with which Flint's periodicity may be meshed algebraically with the eight-fold symmetry of Mendeleev's periodicity invites speculation that (1) on the ionic level some form of simple attractive force, e.g. one related to total atomic-number-equivalent values for hydrated units calculated in accord with Flint's methodology, may underly aspects of chemical behavior; and (2) the helical structure of Flint's periodicity might relate to larger scale natural helical patterns[41-43].

The similar emphasis on the system of 92 naturally-occurring elements in both Flint's and Fuller's periodic structures and the latter's relation to the number of spheres in the third layer of closest-packed spheres (92)[44] draws attention to the coincidence that numbers of spheres in the first and second closest-packed layers (12 and

42, resp.) are identical to Einstein's calculated "Z1" values for "pure gravitational" and "non-symmetric" fields[6b]. (It may also be noted that the terminal number of the natural periodic system [92] falls within the first interval between prime numbers that exceeds five digits [90-96 inclusive] and is at a distance [3 digits] from the previous prime number [89] which is approximately equal to the average of 22 intervals [2.9] between non-consecutive prime numbers through 89.)

And what of the geometry of snowflakes? The open arrangement of atoms in ice crystals argues that hexagonal close-packing is "irrelevant to an explanation of the hexagonal shape of snow crystals"[45]. However, an alternate explanation of how the emerging snowflake might avoid what Kepler described as "the slippery slope into chaos"[46a] may be found in Flint's projection of a role in the formation of water vapor for positively ionized molecular oxygen ($O2+$). Maximally hydrated with a complement of 6 $H2O-$ units[3j], such an $O2+$ ion could release a water sextet when neutralized (presumably through contact with cold air) which would in the first instance, as per Kepler, "be essentially flat, incapable, that is, of combining with itself to form a solid body"[46b]. If frozen in that shape, this could comprise Kepler's "6 points on a circle for 6 prongs to be welded on to them"[46c]. Such an hypothesized role for the shadow of an $O2+$ ion in the formation of a snowflake might be viewed as consistent with Kepler's pondering "whether there is any salt in a snowflake and what kind of salt" which might account for its shape[46d].

(Another of Kepler's musings on the snowflake could similarly be viewed as anticipating the likes of Flint, the suggestion that "Some botanist might well examine the saps of plants to see if any difference there corresponds to the shape of their flowers"[46e]. Three centuries later botanist Flint proposed to integrate studies of aqueous solutions, light sensitivity of plants and wavelengths of absorbed radiation of gases[47-49] in projecting chlorophyll structure[34b].)

Overall, the foregoing may be viewed as consistent with Einstein's desired "purely algebraic theory for the description of reality"[6a], and with Newton's suggestion that "the phenomena of Nature ... may all depend upon certain forces by which the particles of bodies by some causes hitherto unknown, are either mutually impelled towards one another, and cohere in regular figures, or are repelled and recede from one another"[18b]. Spawned in the mainstream of classical scientific thought, Flint's proposed description of hydrational potentiality may come to serve as a fundamental tool in the quest for an understanding of the inner workings of the ponderable universe.

TABLE 1	Equivalent Ionic Conductivities				
Z	C	Obs.(25[0]C)[8] $(10^{-4}m^2$ S $mol^{-1})$	Calculation Error (by Hydrational Assumption)		
			$(Hmax^{2,3})$ $(Hmax/2)$	$(Hmax/4)$	$(Zero^{2,3})$
9 OH	–	198		06	

#	Ion	Value	BASE	Err1	Err2	Err3	Err4
1	H +	349.65					-06
3	Li +	38.66		-03			
4	Be ++	45		08			
9	F -	55.4			-09		
11	Na +	50.08	<BASE>	00			
12	Mg ++	53.0		-02			
13	Al +++	61		03			
17	Cl -	76.31			00		
19	K +	73.48		-05			
20	Ca ++	59.47		-37			
21	Sc +++	64.7				05	
24	Cr +++	67				06	
25	Mn ++	53.5			08		
26	Fe ++	54			07		
26	Fe +++	68				06	
27	Co ++	55			07		
28	Ni ++	50			-04		
29	Cu ++	53.6			01		
30	Zn ++	52.8			-02		
35	Br -	78.1				-02	-14
37	Rb +	77.8					-09
38	Sr ++	59.4		09			
39	Y +++	62		04			
47	Ag +	61.9					-19
48	Cd ++	54			19	-03	
53	I -	76.8					05
55	Cs +	77.2					09
56	Ba ++	63.6					-08
57	La +++	69.7					02
58	Ce +++	69.8					03
59	Pr +++	69.5					04
60	Nd +++	69.4					04
62	Sm +++	68.5					05
63	Eu +++	67.8					04
64	Gd +++	67.3					04
66	Dy +++	65.6					01
67	Ho +++	66.3					05
68	Er +++	65.9					05
69	Tm +++	65.4					05
70	Yb +++	65.6					06
80	Hg ++	63.6					09

Key to Table 1: Hmax = maximum hydration number[2,3]; see equation 2.
Calc. conductivity = k/(Sq. Rt. of Z'h);[2] k=527.62553; see eq. 3.
Error (in percent) = 100([Observed/Calculated] - 1).
TABLE 2   Hydration Values Compared: Gluekauf[36] v.s. Flint[3,4]

Gluekauf Values (X 11/2 =) Adjusted Relative Values

```
H+     3.9                              21  ┐
                                           │
Li+    3.4                              19  │
                                           │  >  = Hmax per Flint
Na+    2.0          <<BASE>>            11  │
                                           │
K+     0.6                               3  │
                                           │
Rb+    0                                 0  │
                                              >  = H value in Table 1
Cs+    0                                 0  ┘
```

TABLE 3a   Solute Ionic Diameters & Equivalent Conductivities

				Diameter (Relative to Na+)			Conductivity ($10^{-4}m^2$ S mol^{-1})		
Z	C		Hmax	Ass.	Obs.[37]	E	Ass.	Obs.[17]	E
17	Cl	−	7	Hmax/2	.65	−08	Hmax/2	76.3	00
11	Na	+ <BASE>	11	Hmax	1.00	00	Hmax	50.1	00
19	K	+	3	Hmax	.68	−02	Hmax	73.5	−05
31	HCO3	−	16	Hmax	1.12	−01	Hmax	44.5	11
31	CH3COO	−	16	Hmax	1.22	07	Hmax	40.9	02
49	H2PO4	−	21	Hmax	1.39	11	Hmax	33	−04
48	SO4	−−	23	Hmax	1.25	−03	0	80	03
48	HPO4	−−	23				Hmax	33	−01
	2(HPO4	−−)		2Hmax	1.76	08			

TABLE 3b   Crystal Ionic Radii & Solute Equivalent Conductivities

				Ionic Radius (10^{-10}m)			Conductivity ($10^{-4}m^2$ S mol^{-1})		
Z	C		Hmax	Ass.	Obs.[38]	E	Ass.	Obs.[17]	E
3	Li	+	19	0	.60	−09	Hmax	38.66	−03
11	Na	+ <BASE>	11	0	.95	00	Hmax	50.08	00
19	K	+	3	0	1.33	18	Hmax	73.48	−05
37	Rb	+	8	0	1.48	06	0	77.8	−09
55	Cs	+	13	0	1.69	06	0	77.2	09

Hmax = "Maximum" hydration number per Flint, as per equation 2.
Ass. = Hydrational assumption
Obs. = Observed
E = % Error = 100([Observed/Calculated] − 1)
    Diameter/radius = cube root of relative VZ; see equation 4.
    Conductivity = inverse-sq.-rt. of relative Z'h; see eq. 3.

Key to Table 3a and 3b:

Hmax = "Maximum" hydration number per Flint, as per equation 2.
Ass. = Hydrational assumption

Obs. = Observed
E = % Error = 100([Observed/Calculated] - 1)
    Diameter/radius calculated as cube root of relative VZ;
        VZ calculated as per equation 4.
    Conductivity calculated as inverse-sq.-rt. of relative Z'h;
        Z'h calculated as per equation 3.

TABLE 4   Energy of Solution Components (kcal/mol)[38];
    approximated as Relative Square-Root of Z'h values.

	/\Esol[38a]		ESX1[38b]		ESX[38b]		ESX2[38b]	
	Obs.	Error	Obs.	Error	Obs.	Error	Obs.	Error
Li+	-165 ± 4	05	-138	-08	-231	-06	-92	-04
Na+ <BASE>	-125 ± 2	00	-119	00	-195	00	-76	00
F-	-111 ± 4	08#	-129	-04	-212	-04	-83	-04
Cl-	-80 ± 5	-02#	-77	-01# -23	-143	-13	-66	03

Key to Table 4:

Error = 100([Observed Energy Component[38]]/[Approximation]) -1;
    Approximation = Sq. Rt. of Z'h, relative to Na+ (BASE)
        = (Sq.Rt. of [Z'h/111])(Obs. value for Na+);
        Z'h calculated as per equation 3[3].
"#" - Denotes calculation based on half-maximum hydration level;
        all   other   calculations   based   on   maximum   hydration   level[3].

TABLE 5   Hydrated Atomic-Number-Equivalents (Z'h) per Flint[3]
    v.s. Mendeleev Periodicity

1	2	3	4	5	Mendeleev
Z'h (Z)	Z'h (Z)	Z'h (Z)	Z'h (Z)	Z'h (Z)	Group
		148(85)	212(77)	276(69)	
	85(67)	149(59)	213(51)		------------ (V)
	86(**41**)	150(33)	214(25)		
23(23)	87(15)	**151(7)**			
	92(92)	156(84)	220(76)		
	93(66)	157(58)	221(50)		------------ (IV)
	94(40)	158(32)	222(24)		
31(22)	95(14)	159(6)			

```
                    100(91)      164(83)      228(75)

                    101(65)      165(57)      229(49)  ------------(III)

                    102(39)      166(31)      230(23)

        39(21)      103(13)      167( 5)

                    108(90)      172(82)      236(74)

                    109(64)      173(56)      237(48)  ------------(II)

        46(46)      110(38)      174(30)

        47(20)      111(12)      175( 4)

                    116(89)      180(81)      244(73)

                    117(63)      181(55)      245(47)  ------------(I)

        54(45)      118(37)      182(29)

        55(19)      119(11)      183( 3)

                    124(88)      188(80)      252(72)

                    125(62)      189(54)      253(46)

        62(44)      126(36)      190(28)

        63(18)      127(10)      191( 2)  -----------------------(0)

                    132(87)      196(79)      260(71)

        69(69)      133(61)      197(53)  -----------------------(VII)

        70(43)      134(35)      198(27)

        71(17)      135( 9)      199( 1)

                    140(86)      204(78)      268(70)

        77(68)      141(60)      205(52)  -----------------------(VI)

        78(42)      142(34)      206(26)

        79(16)      143( 8)      207( 0)
```

Key to Table 5:
Z = Atomic number
Z'h = Z + 9(Hmax); Hmax calculated as per equation 2, after Flint[3].
Bold numbers denote Z values through Z=57 which are prime numbers and Z'h values paired with these Z values which are also primes.

References:

1. Einstein, A., letter to Max Born, 7 September 1944, in <u>Einstein,A Centenary Volume</u> (ed French, A.P.), 276 (Harvard U. Press, Cambridge, 1979).

2. Flint, L. H., <u>J. Wash. Acad. of Sci.</u> **22**, 97-119, 211-217 & 233-   237 (1932).

3. Flint, L. H., <u>Behavior Patterns of Hydration</u> (Institute for Advancement of Science and Culture, New Delhi, 1964), (a) 20-22, (b) 16, (c) 18, (d) 19, (e) 25, (f) 80, 92, (g) 130-8, (h) 30ff, (i) 39ff, (j) 119.

4. Weingärtner, H. & C. A. Chatzidmitriou-Dreismann, <u>Nature</u> **346**, 547-550 (1990).

5. Eigen, M. and DeMaeyer, L., <u>Proc. R. Soc.</u> **A247**, 505-533 (1958),   p. 509.

6. Einstein, A., (Dec.1954) <u>Meaning of Relativity</u>, 5th Ed. (Princeton U. Press, Princeton, 1956), (a) 160-1, (b) 166.

7. Maddox, J., <u>Nature</u> **347**, 13 (1990).

8. <u>CRC Handb. of Chem. and Phys.</u>, D167-8 (Chemical Rubber Co., Cleveland, Ohio, 1985-6).

9. Arrhenius, S., <u>Theories of Chemistry</u>, 91 (Longmans, Green and Co., London, 1907).

10. Bredig, G., <u>Zeitschr. Phys. Chem.</u> **13**, 262 (1894).

11. Robinson, Frank N. H., in <u>Encyclopedia Britannica</u> 15th Edition   (1988), Vol. 18, p.268.

12. Abegg and Bodländer, <u>Zeit. f. Anorg. Chem.</u> 454-499 (1899).

13. Senter, G., <u>Trans. Faraday Soc.</u> **3**, 146-152 (1907).

14. Bayliss, W. M., <u>Principles of General Physiology</u>, 177-8 (Longmans, Green and Co., London, 1915).

15. Nernst, Walter, <u>Theoretical Chemistry</u>, 397 (Macmillan and Co., Ltd., London, 1916).

16. Perlzweig, W. A., <u>Principles of the Theory</u>, in Leonor Michaelis, <u>Hydrogen Ion Concentration</u>, 125 (Baltimore, Williams and Wilkins Co., 1926).

17. Graham, Thomas, <u>Proc. Roy. Soc.</u> **12**, 611-623 (1863), p. 614.

18. Pauling, L., <u>Chemical Bond</u>, 174 (Cornell U. Press, Ithaca, N.Y., 1967).

19. Newton, I., <u>Principia</u> (1687); (a) II, Prop.33; (b) (1686), Preface to 1st Edition.

20. van't Hoff, J. H., <u>Zeitschr. f. Phys. Chem.</u> **i**, 481-508 (1887) in <u>Alembic Club #19</u> (Alembic Club, Edinburgh, 1929).

21. Kohlrausch, F., <u>Gottingen Nachrichten</u>, 213 (1876) in Goodwin, <u>Scientific Memoirs</u> (MIT, Boston).

22. Einstein, A., <u>Zeit. f. Elektrochem.</u> **14**, 235-9 (1908), per<u>Theory of the Brownian Movement</u>, 84-5 (Dover, 1956).

23. Einstein, A., <u>Theory of the Brownian Movement</u>, Preface (Dover, 1956).

24. Einstein, A., Zeit. f. Elektrochemie **13**, 41-2 (1907), perBrownian Movement, 64 (Dover, 1956).

25. Einstein, A. (1942), in Out of my Later Years, 233 (Philosophical Library, New York, 1950).

26. Bousfield, W. R., Proc. Royal Soc. **88A**, 147-169 (1912).

27. Encyclopedia Britannica, 15th Ed. (1991), Vol.1, p. 676.

28. Moseley, H. G. J., Phil. Mag., 703 (1914).

29. Bertsch, G. F. and S. McGrayne, in Encyplopedia Britannica, 15th Ed. (1991), Vol. 14, p. 330.

30. Bousfield, W. R., Phil. Trans. **206A**, 101-150 (1906), p.124.

31. Flint, L. H., Plant Physiology **9**, 107-126 (1933).

32. Amiss, E. S. and J. F. Hinton, Solvent Effects on Chemical Phenomena (Academic Press, New York and London, 1973), 52-54 (Table 3.1), 65-67 (Table 3.4), 109-110 (Table 3.15).

33. Gluekauf, E., Faraday Soc., Transactions **51** 1241 (1955).

34. Flint, L. H., Hydration and Biology, 104ff (Institute for the Advancement of Science and Culture, New Delhi, 1968), (a) 104ff, (b) 45ff.

35. Flint, L. H., Dissenting Ape, 25-27 (Carlton, N. Y., 1973), (a) 53, (b) 25-27.

36. Ganong, W. F., Review of Medical Physiology 12 (Lange, Los Altos, 1975).

37. Hildebrand, J. H., Reference Book of Inorg. Chem., Revised Ed. 37 (Macmillan, N.Y., 1940).

38. Chandrasekhar, J., D. C. Spellmeyer and W. L. Jorgensen, J. Am. Chem. Soc. **106**, 903-910 (1984), (a) p. 905 (Table 2), (b) 908 (Table 4).

39. Pauling, L., Encyclopedia Britannica, 15th Ed. (1991), Vol. 15, p. 938.

40. Fuller, R. B., Synergetics 2 (MacMillan, N. Y., 1979), 403-4.

41. Amato, J., Science **255**, 255-6 (1992).

42. Shibata, K. and R. Matsumoto, Nature **353**, 633-5 (1991).

43. Watson, J. D. and F. H. C. Crick, Nature **171**, 737-8 (1953).

44. Fuller, R. B., Synergetics (MacMillan, N. Y., 1975), 133.

45. Mason, B. J. in Kepler, J. (1611), A New Years' Gift or On The Six-Cornered Snowflake 54 (Clarendon Press, Oxford, 1966).

46. Kepler, J. (1611), A New Years' Gift or On the Six-Cornered Snowflake (Clarendon Press, Oxford, 1966), (a) 29, (b) 41, (c) 23, (d) 45, (e) 43.

47. Flint, L. F., Science **80**, 38-40 (1934).

48. Flint, L. F., and McAlister, E. D., Smithsonian Miscellaneous Collections **94**, No. 5, 1-11 (Publication 3334, 14 June 1935); and **96**, No. 2, 1-9 (Publication 3414, 16 June 1937).

49. Flint, L. F., and Moreland, C. F., Am. J. Botany **26**, 231-3 (1939).

- From Lindley, Nature, 21 Aug. 1992
"In reply please quote:SO8279 DL/kb

August 21, 1992

Dr. S.H. Shakman
PO Box 382
Santa Monica, CA 90406-0382

Dear Dr. Shakman,

Thank you for your manuscript "Conductance, hydration and periodicity". Unfortunately we cannot offer to publish it in <u>Nature</u>.

Your paper is far too long to be published in <u>Nature</u>, and in any case seems to be largely a historical account of a specialized debate, which does not seem likely to be of interest to a wide range of our readers. Frankly, your final sentence, which suggests a larger importance to this work, seems quite unsupported by anything that has gone before.

I am sorry we must be negative.

Yours sincerely,
Dr. David Lindley
Associate Editor

- To Lindley, 24 December 1992

> P.O. Box 382
> Santa Monica, CA 90406-0382
> (310)453-7707
> 24 December 1992

Dr. David Lindley
<u>Nature</u>
1234 National Press Building
Washington, D.C. 20045

Reference: S08279 DL/kb

Dear Dr. Lindley,

  Thank you for your kind and helpful note of 21 August 1992 accompanying the return of "Conductance, hydration and periodicity" (#S08279).
  Enclosed is a revised and shortened submission entitled "Not just a flake model: hydration - conductance to quadruple helix", which seeks to expose and extend the potential utility of L. H.

Flint's hydration methodology.

The proposed central significance of this work relates to its potential contribution to understanding the nature of water and aqueous solutions, which C. M. Sorensen recently referred to in <u>Nature</u> (**360**, 303 [26 November 1992]) as "the most far reaching of solvents" and "grand mediator of ... biochemistry."

Further possible relevance to <u>Nature</u> readers may, for example, be found in Flint's studies of light sensitivity of plants, relative to his hydration work and radiation absorption of gases (Refs. 14-16 and 11b of enclosed), which studies might usefully augment perspectives recently discussed by E.M. Meyerowitz in <u>Nature</u> (**360**, 419 [3 December 1992]).

Possible broader implications might follow from circumstances that Flint's methodology, for example, is algebraic, though not through any prior intention of Flint; and facilitates consideration of chemical and proposed hydrational periodicity schemes, and patterns of alignment of prime numbers therein, within a common framework. Opening and closing statements of my previous submission (S08279), which had sought to reflect such possible broader implications, have been replaced in the enclosed by ones which more directly refer to the paper's proposed central significance.

Thank you for any possible consideration.

Wishing you a happy holiday season and healthy new year,

Sincerely yours,
Stuart Hale Shakman

enclosures

___

The following artice was submitted with above 24 Dec. letter

PROPOSED ARTICLE

Not just a flake model: hydration - conductance to quadruple helix

Stuart Hale Shakman

P. O. Box 382, Santa Monica, CA, U. S. A. 90406-0382

"MODEL" qualities of calculability and comprehensibility[1,2] are to be found in Flint's purported quantification of the venerable hydrate theory[3,4]. Flint's work itself originated in calculations from conductance data via Graham's law, and allows for calculating various phenomena from first principles within a comprehensible (quadruple helical, periodic) framework. Original correlations discussed herein address an extended range of conductivities, alternately-determined hydration numbers, linear size (along with conductance) for two data sets, heats of hydration, snowflake structure, and chemical periodicity.

For more than a century, the hydrate theory of solutions has remained a viable explanation of conductance and other solute phenomena[5,6]. But despite the importance of aqueous solution dynamics to many physical, chemical and biological processes[7], agreeing on precise numbers of water-solvent units associated with various solute ions continues to be a problem.[8,9] In 1932 Flint uncovered the basis for a proposed solution by calculating proposed hydrated weight values[4] from relative conductance, in accord with Graham's law[10]. Flint's calculations disclosed an apparent pattern of hydration which he found amenable to algebraic description. His subsequent efforts to validate or otherwise extend the potential utility of this so-named "description of hydrational potentiality" included reference to numerous phenomena including specific gravity[3a], osmosis[3b], gas solubility[11a], densities of elemental solids[12a], acidity-alkalinity[3c,13] and radiation sensitivity[14-16] in conjunction with chlorophyll structure[11b].

This paper seeks to extend the use of Flint's methodology through correlations with: relative conductance for a wide range of ions[17] (Table 1) which in part support an 1899 suggestion[18] that high conductivities of H+ and OH- ions are due to relatively low weight (independent of consideration of proposed quantum effects[7]); other determinations of relative hydration values (e.g. as in Table 2); both linear size and conductivities for two sets of ions (Table 3); four component sets of energies of solution (Table 4); and the six-cornered snowflake[19]. In addition, correlations with Mendeleev's groups (Table 5) may be viewed as suggesting a possible gravitational contribution to underlying chemical behavior; and Flint's 24-by-4 periodic matrix (Fig. 1), wrapped into a quadruple helix, dovetails with R. B. Fuller's work[20] and also constitutes a three-dimensional framework for illustrating chemical periodicity[21].

Background

The notion that distinct numerical combinations of water and solute units exist within aqueous solutions, the so-called hydrate theory of solutions, is closely associated historically with Dimitri Mendeleev's early (1887) and strong advocacy[5]. Bredig's subsequent use of this concept in 1894 to explain why the chlorine ion is less mobile than the heavier iodine ion, attributing this to a larger hydration shell around the Cl- ion that migrates with it[22], has remained essentially unchanged up to the present[6].

In 1899 Abegg and Bodländer sought to explain why mobility of the K+ ion is greater than that of the smaller and lighter Na+ ion, and the latter greater than that of the even smaller and lighter Li+ ion, with the suggestion that hydration potential may vary inversely with anhydrous weight[18]. This could allow for the hydrated Li+ ion to be heaviest and least mobile of the three ions, and the much-less-hydrated K+ ion to be lightest and most mobile. In accord with this same premise, large conductivities of H+ and OH- were attributed to an absence of hydration.

Over the next few decades, other investigators including the renowned Walter Nernst[23-26] similarly invoked the concept of hydration in discussions of conductivities; however, a mathematical framework was not advanced until 1932, when Flint called on Graham's law and in-so-doing exposed an apparent algebraic relationship between solute ions and their water solvent. As per Graham and as required by the more recent law of kinetic energy[27], the mobility of a gas varies inversely with the square root of its (density or) molecular weight. As solute behavior had been shown by van't Hoff to be analogous to gaseous[28], and conductivity was recognized as an index of mobility[29,30], Flint was able to view conductance as a measure (inverse-square-root) of relative hydrated weights of solute ions.

In his initial work,[4] Flint cited Nernst conductivity values for K+, Na+ and Li+ of 65.3, 44.4, and 35.5, resp., as listed W. Bayliss's 1915 physiology textbook[26]. (On an adjoining page of this text were listed W. Bousfield's proposed hydration numbers 4, 8 and 16 for these same ions, a self-proclaimed "attractive looking series"[31].) Inspired by Abegg and Bodländer[3d], Flint sought to determine hydration numbers that would allow for an optimum fit with the Nernst data, via Graham's law. He found that presumed hydration numbers of 4, 12, and 20, each multiplied by 18 (for the weight of each water unit) and added to conventionally-appraised atomic weight values of (approx.) 39 for potassium (K), 23 for sodium (Na) and 7 for lithium (Li), resp., yielded totals (111, 239 and 367) whose relative inverse-square-roots (.0949, .0647 and .0522) correlated remarkably well with the Nernst data.

As his initially-hypothesized hydration numbers (4, 12, and 20) were integers, each of which when added to its associated atomic number (19, 11 and 3 for K, Na and Li, resp.) equalled 23, Flint attempted to use integral values based on atomic numbers in his conductivity calculations[3e]. He then found that adjusted atomic-number-equivalent values (Z') which incorporated a shift equal to one unit Z per unit valence (C) "permitted an integration of observational data with a satisfying convincing nicety"[3f] (equation 1):

$$Z' = Z+C, \qquad\qquad\qquad (1)$$

where Z=atomic number and C=valence. (Flint's use of values based on atomic numbers is consistent with contemporary recognition of the primacy of the atomic number[32] as established by Moseley[33]; contemporary science also embraces the concept of a shift in atomic-number-equivalent values in the case of an increase in the atomic number of some radioactive elements resulting from the loss of a nuclear electron[34].)

Noting that the number 23 equals one-fourth of the number (92) of naturally-occurring elements, Flint projected four equal periods, in each of which the maximum hydration number (Hmax) decreases from 23 to zero with increasing atomic number[3g,4] (equation 2):

$$\text{Hmax} = 23n - Z', \text{ when Hmax} = 23 \text{ to } 0, \text{ and } n = 1 \text{ to } 4 \quad (2)$$

$$(\text{for } Z' = 0 \text{ to } 23, n=1;$$
$$\text{for } Z' = 23 \text{ to } 46, n=2;$$
$$\text{for } Z' = 46 \text{ to } 69, n=3;$$
$$\text{for } Z' = 69 \text{ to } 92, n=4).$$

In his calculations Flint characterized water units involved in hydration as negatively-charged ions (H_2O-)[3g], allowing for calculation of the relative hydrated Z' value (Z'h) for a given solute ion as the sum of Z' for the ion plus a value of 9 for each associated H_2O- unit (equation 3)[3h]:

$$Z'h = Z' + 9H, \qquad\qquad\qquad (3)$$

where H is the actual number of associated H_2O- units.

Approximations of ionic data

Table 1 displays observed conductivities for the 41 monatomic elemental ions through Z=80, plus OH-, listed in CRC[17] and error values (E) for approximations derived as the relative inverse- square-root of Z'h, under one of four hydrational assumptions - maximum, zero, half-maximum and one-fourth-maximum hydration levels. Flint had primarily utilized assumptions of either maximum or zero hydration, which assumptions alone notably enable conductivity calculations with error values of 10% or less for all of 22 CRC-listed[11] mono-atomic positively-charged elemental ions with Z = 1-19 or 55-80, plus OH-. Calculations for both H+ and OH- involve the assumption of zero hydration.

Flint also introduced the concept of shared water ions within discussions of hydrational bonding and utilized fractional-maximum-hydration levels in associated calculations[3i]. The specific impetus for the use herein of a speculative half-maximum-hydration assumption is the observation that, as measured against presumed maximum hydration for the base ion Na+, calculations of both conductivity and diameter for the Cl- ion are optimum using an hydration number of 3.5 (as in Table 3a), a value first proposed by Bousfield in 1906[35]. It was noted that this value is equal to half the maximum hydration number as per Flint, and further that a half-maximum-hydration assumption also allows for agreeable conductivity calculations for F- and for the 6 divalent "transition-series" ions Mn++, Fe++, Co++, Ni++, Cu++ and Zn++. Thus conductivities for all but 7 of the 42 ions in Table 1 are within 10% of calculated values, assuming either maximum, zero or half-maximum hydration levels. A single fourth (speculative) assumption of a one-fourth-maximum hydration level allows for approximations of conductivities for 5 of these 7 with error values less than 10%. (Note clustering of distinct numerical groupings under each hydrational assumption.)

Calculations relating to relative mobilities of ions, aside from playing an indispensable role in Flint's initial work on hydration, were subsequently used by Flint in discussions of the concept of "Ph" and development of a "tentative" acidity-alkalinity index[3c,13].

Although various efforts to determine hydration numbers have yielded a wide range of values, the ratio of values for Na+ v.s. Li+ per Flint (11/19 = .58) correlates reasonably well with the average (.61) for 14 sets of values by 9 investigators as listed by Amiss[9]. Even more striking is how Gluekauf's relative hydration values[36] for the series H+, Li+, Na+, K+, Rb+ and Cs+, adjusted to a base value of 11 for Na+ as in Table 2, correlate precisely with Flint's maximum (H+, Li+, Na+ and K+) or zero (Rb+ and Cs+) hydration levels; except for H+, all adjusted values in Table 2 are identical with hydration values used in Table 1.

Seeking to validate his work through approximation of specific gravities, Flint proposed that relative volume of a solute ion (VZ) may be approximated as the quotient of (a) Z'h and (b) 1 + Z'/Z'h (equation 4)[3a]:

$$VZ = Z'h/(1+ Z'/Z'h) \qquad (4)$$

Flint used this formulation in deriving theoretical values for comparison with specific gravity data[3a], results of original osmosis experiments[3b], gas solubility data[11a], and densities of elemental solids[12a], but is not known to have used it in conjunction with linear ionic size.

Table 3a lists observed relative diameter values for a set of 8 solute ions of physiological interest[37], observed conductivities, and approximation-error values for both sets of data. Diameters are approximated as cube-root of VZ, relative to a base value of 1.00 for Na+; conductivities are approximated as inverse-square-root of Z'h, relative to 50.08 for Na+ as in Table 1.

As shown in Table 3a, observed diameters for 7 of the 8 listed ions (all except HPO4--), and observed conductivities for all 8 ions, are within 11% of calculated values, assuming a "maximum" hydration level in all cases except (a) a half-maximum-hydration assumption used in calculations of both diameter and conductivity of Cl-, and (b) a zero hydration assumption used in calculation of conductivity of SO4--. In the case of HPO4--, conductivity has been approximated using the assumption of maximum hydration, but diameter may be approximated (as relative cube-root of VZ) only if ionic size is assumed to be even larger -- approximately equal to that of two fully hydrated ions (which hypothetical "double" ion might be projected as splitting under electrical stress).

Table 3b lists observed crystal ionic radii[38] and approximation error values for the series Li+, Na+, K+, Rb+ and Cs+. Radii are approximated as relative cube root of theoretical volume (VZ), as were diameters in Table 3a, except that for all ions in Table 3b approximations of radii involve the assumption of zero hydration. While the order of agreement for approximations of linear size is less satisfactory in Table 3b than in Table 3a, it may be noted that the trend in error values in Table 3b for Li+, Na+ and K+ radii is consistent with Flint's discussion of how the attractive force mediating hydration may also act as a relative cementing force in solids[12a]; this might allow for the Li+ ion, with a greater hydrational/attractive potential than Na+, to be more compressed in crystal form, and the K+ ion, with a lesser potential than Na+, to be less compressed.

Hydrational potentiality per Flint is correlated with components of energies of solution[39] in Table 4. Relative square-roots of Z'h values associated with maximum hydration for Li+ and Na+ against half-maximum hydration for F- and Cl- (assumptions used in Table 1) correlate reasonably well with "total solvent energy" values (_Esol); whereas relative square-roots of Z'h values associated with "maximum" hydration for all 4 ions yield comparably-satisfactory approximations of the "bulk solvent" portion (ESX2) but not first shell portion (ESX1) of total solute-solvent energy (ESX; ESX = ESX1 + ESX2). For the first shell (ESX1) a shift in assumptions for the Cl- ion, from a maximum to a half-maximum hydration level, yields an improved correlation for Cl- relative to calculations based on the assumption of full hydration for the other three ions.

The snowflake

While the open arrangement of atoms in ice crystals argues that hexagonal close-packing is "irrelevant to ... the hexagonal shape of snow crystals"[40], an alternate to this or other possible explanations[19] of how the emerging snowflake might avoid "the slippery slope into chaos"[41a] may be found in Flint's projection of a role in the formation of water vapor for positively ionized molecular oxygen (O2+). Maximally hydrated with a complement of 6 H2O- units[3j], such an O2+ ion could be projected as releasing a water sextet when neutralized which would in the first instance, as per Kepler, "be essentially flat, incapable, that is, of combining with itself to form a solid body"[41b]. If frozen in that shape, this could comprise Kepler's "6 points on a circle for 6 prongs to be welded on to them"[41c]. Such an hypothesized role for the shadow of an O2+ ion in the formation of a snowflake might be viewed as consistent with Kepler's pondering "whether there is any salt in a snowflake and what kind of salt" which might account for its shape[41d].

Periodicity and the quadruple helix

Prospective correlations between Flint's and Mendeleev's periodic schemes are illustrated in Table 5 and Figure 1. Table 5 exhibits in ascending order Z'h values for all values of Z from one through 92. These Z'h values fall into groupings of up to four consecutive values (e.g., 148, 149, 150, 151), each value 8 distant from the same position in the next grouping (156, 157, 158, 159). Data in Table 5 is displayed in (five) columns with a difference between Z'h values horizontally of 64, which arrangement allows all atomic numbers from Z=2 through Z=57 that fall within Mendeleev's Groups 0 through VII, except Z=24 & 25, to fall into corresponding horizontal groupings; the Z value for hydrogen (Z=1) is found with Z values for elements in Mendeleev's Group VII, where hydrogen is sometimes categorized due to some chemical similarities[42]. (Unique structural characteristics of each of two adjacent horizontal groupings may be noted: [a] Within that corresponding to Mendeleev's Group 0, Z'h for Z=18 is 63, and all other Z'h values are approximate multiples; and [b] in the grouping corresponding to Mendeleev's Group 1, all Z values through Z=47 are prime numbers, none of which is paired with a prime-numbered Z'h.)

In Figure 1 Mendelev's groups are projected on a 24 X 4 matrix representing Flint's periodicity. This matrix (including "0" and duplicates of "23", "46", and "69") bears an uncanny resemblance to R. B. Fuller's "four of the 24-ness of the duo-tet cube" (including a "disappearing octa set" of one "expendable exterior octa" and "three expendable interior octa")[20]. A prospective "disappearing octa set" does indeed virtually disappear when Flint's periodic matrix is characterized continuously and three-dimensionally as a quadruple helix. Figure 1 may thus be viewed as a comprising a two-dimensional portrayal of such a quadruple helix, encompassing somewhat symmetrical graphic representations of "nonmetals", "lanthanides" and "actinides".

Conclusions and speculations

Beyond arguing for consideration of Flint's methodology as may relate to a general understanding of solute phenomena, this paper specifically supports (1) Bousfield's hypothesis of an hydration number of 3.5 for Cl-[35] (incident to measurement of conductivity and diameter as in Table 2), offering that this may represent a level equal to one-half the maximum hydration level per Flint and further that conductivities of several other ions may similarly indicate hydration levels which may be derived as distinct fractions of respective maximum (per Flint) hydration levels; (2) the hypothesis that the sulfate ion, fully hydrated when measured for diameter, may shed its water complement under electrical stress incident to measurement of conductivity; and (3) the work of Abegg and Bodländer[18], and others[22,25,4], suggesting that large conductivities of OH- and H+ ions in solution are due to a relative absence of hydration, although these ions may otherwise be strongly hydrated as has been inferred from studies of specific gravities of acidic aqueous solutions[12b] and relatively large heats of hydration[43]. Other possible shifts in hydrational states also evidenced in above-discussed calculations of size and conductance (Table 3) and energies of solution (Table 4).

Beyond consideration of the shape of snowflakes, Kepler might further be viewed as anticipating the likes of Flint with the suggestion that "Some botanist might well examine the saps of plants to see if any difference there corresponds to the shape of their flowers"[41e]. Three centuries later botanist Flint proposed to integrate studies of aqueous solutions, light sensitivity of plants and wavelengths of absorbed radiation of gases[14-16] in projecting chlorophyll structure[11b].

The ease with which Flint's periodicity may be meshed algebraically with Mendeleev's (Table 5) argues against a spurious origin for the latter[21] and may prompt speculation that some simple attractive force, e.g. one related to total atomic-number-equivalent values for hydrated units calculated in accord with Flint's methodology, may underlie aspects of chemical behavior. Quadruple helical portrayal of Flint's 24-by-4 periodicity (Fig. 1), beyond comprising a proposed three dimensional array of chemical elements, might be speculatively: related to larger-scale natural helical patterns[44-46]; and somehow defined by a somewhat symmetrical distribution of prime numbers within (89+1 = 83+7 = 79+11, etc.) and terminal point (92) which falls within the first interval between primes exceeding five numbers (90-96 inclusive).

The similar emphasis on the system of 92 naturally-occurring elements in both Flint's and Fuller's periodic structures and the latter's relation to the number of spheres in the third layer of closest-packed spheres (92)[47] draws attention to the coincidence that numbers of spheres in first and second closest-packed layers (12 and 42, resp.[47,48]) are identical to Einstein's calculated "Z1" values for "pure gravitational" and "non-symmetric" fields[49].

Overall the foregoing argues for consideration of Flint's methodology as a proposed characterization of aqueous solution and other phenomena with desirable qualities beyond those expected of otherwise virtuous intermediate models[2].

TABLE 1   Equivalent Ionic Conductivities

Z	C	Obs.(25°C)[8] $(10^{-4} m^2 \ S \ mol^{-1})$	Calculation Error (by Hydrational Assumption)			
			(Hmax[2,3])	(Hmax/2)	(Hmax/4)	(Zero[2,3])
9	OH −	198				06
1	H +	349.65				−06
3	Li +	38.66	−03			
4	Be ++	45	08			
9	F −	55.4		−09		
11	Na +	50.08 <BASE>	00			
12	Mg ++	53.0	−02			
13	Al +++	61	03			
17	Cl −	76.31		00		
19	K +	73.48	−05			
20	Ca ++	59.47	−37			
21	Sc +++	64.7			05	
24	Cr +++	67			06	
25	Mn ++	53.5		08		
26	Fe ++	54		07		
26	Fe +++	68			06	
27	Co ++	55		07		
28	Ni ++	50		−04		
29	Cu ++	53.6		01		
30	Zn ++	52.8		−02		
35	Br −	78.1			−02	−14
37	Rb +	77.8				−09
38	Sr ++	59.4	09			
39	Y +++	62	04			
47	Ag +	61.9				−19
48	Cd ++	54		19	−03	
53	I −	76.8				05
55	Cs +	77.2				09
56	Ba ++	63.6				−08
57	La +++	69.7				02
58	Ce +++	69.8				03
59	Pr +++	69.5				04
60	Nd +++	69.4				04
62	Sm +++	68.5				05
63	Eu +++	67.8				04
64	Gd +++	67.3				04
66	Dy +++	65.6				01
67	Ho +++	66.3				05
68	Er +++	65.9				05
69	Tm +++	65.4				05
70	Yb +++	65.6				06
80	Hg ++	63.6				09

Key to Table 1: Hmax   = maximum hydration number[2,3]; see equation 2.
Calc. conductivity = k/(Sq. Rt. of Z'h);[2] k=527.62553; see eq. 3.
Error (in percent) = 100([Observed/Calculated] - 1).

TABLE 2   Hydration Values Compared: Gluekauf[36] v.s. Flint[3,4]

Gluekauf Values (X 11/2 =) Adjusted Relative Values

H+	3.9		21
Li+	3.4		19
Na+	2.0	<<BASE>>	11
K+	0.6		3
Rb+	0		0
Cs+	0		0

> = Hmax per Flint

> = H value in Table 1

TABLE 3a   Solute Ionic Diameters & Equivalent Conductivities

Z		C		Hmax	Diameter (Relative to Na+)			Conductivity ($10^{-4}m^2$ S mol^{-1})		
					Ass.	Obs.[37]	E	Ass.	Obs.[17]	E
17	Cl	-		7	Hmax/2	.65	-08	Hmax/2	76.3	00
11	Na	+	<BASE>	11	Hmax	1.00	00	Hmax	50.1	00
19	K	+		3	Hmax	.68	-02	Hmax	73.5	-05
31	HCO3	-		16	Hmax	1.12	-01	Hmax	44.5	11
31	CH3COO	-		16	Hmax	1.22	07	Hmax	40.9	02
49	H2PO4	-		21	Hmax	1.39	11	Hmax	33	-04
48	SO4	--		23	Hmax	1.25	-03	0	80	03
48	HPO4	--		23				Hmax	33	-01
	2(HPO4	--)			2Hmax	1.76	08			

TABLE 3b   Crystal Ionic Radii & Solute Equivalent Conductivities

Z		C		Hmax	Ionic Radius ($10^{-10}m$)			Conductivity ($10^{-4}m^2$ S mol^{-1})		
					Ass.	Obs.[38]	E	Ass.	Obs.[17]	E
3	Li	+		19	0	.60	-09	Hmax	38.66	-03
11	Na	+	<BASE>	11	0	.95	00	Hmax	50.08	00
19	K	+		3	0	1.33	18	Hmax	73.48	-05
37	Rb	+		8	0	1.48	06	0	77.8	-09
55	Cs	+		13	0	1.69	06	0	77.2	09

Hmax = "Maximum" hydration number per Flint, as per equation 2.
Ass. = Hydrational assumption
Obs. = Observed
E   = % Error = 100([Observed/Calculated] - 1)
    Diameter/radius = cube root of relative VZ; see equation 4.
    Conductivity = inverse-sq.-rt. of relative Z'h; see eq. 3.

Key to Table 3a and 3b:

Hmax = "Maximum" hydration number per Flint, as per equation 2.
Ass. = Hydrational assumption
Obs. = Observed
E = % Error = 100([Observed/Calculated] - 1)
    Diameter/radius calculated as cube root of relative VZ;
        VZ calculated as per equation 4.
    Conductivity calculated as inverse-sq.-rt. of relative Z'h;
        Z'h calculated as per equation 3.

TABLE 4   Energy of Solution Components (kcal/mol) [38];
           approximated as Relative Square-Root of Z'h values.

	/\Esol[38a]		ESX1[38b]		ESX[38b]		ESX2[38b]	
	Obs.	Error	Obs.	Error	Obs.	Error	Obs.	Error
Li+	−165±4	05	−138	−08	−231	−06	−92	−04
Na+ <BASE>	−125±2	00	−119	00	−195	00	−76	00
F−	−111±4	08#	−129	−04	−212	−04	−83	−04
Cl−	−80±5	−02#	−77	−01# −23	−143	−13	−66	03

Key to Table 4:

Error = 100([Observed Energy Component[38]]/[Approximation]) -1;
    Approximation = Sq. Rt. of Z'h, relative to Na+ (BASE)
        = (Sq.Rt. of [Z'h/111])(Obs. value for Na+);
            Z'h calculated as per equation 3^3.
"#" - Denotes calculation based on half-maximum hydration level;
    all other calculations based on maximum hydration level[3]

TABLE 5   Hydrated Atomic-Number-Equivalents (Z'h) per Flint[3]
                    v.s. Mendeleev Periodicity

1 Z'h (Z)	2 Z'h (Z)	3 Z'h (Z)	4 Z'h (Z)	5 Z'h (Z)	Mendeleev Group
		148 (85)	212 (77)	276 (69)	
	85 (67)	149 (59)	213 (51)	------------ (V)	
	86 (**41**)	150 (33)	214 (25)		
23 (23)	87 (15)	**151 (7)**			
	92 (92)	156 (84)	220 (76)		

	93(66)	157(58)	221(50) ------------(IV)
	94(40)	158(32)	222(24)
31(22)	95(14)	159(6)	
	100(91)	164(83)	228(75)
	101(65)	165(57)	229(49) ------------(III)
	102(39)	166(**31**)	230(23)
39(21)	**103**(**13**)	**167**(**5**)	
	108(90)	172(82)	236(74)
	109(64)	173(56)	237(48) ------------(II)
46(46)	110(38)	174(30)	
47(20)	111(12)	175(4)	
	116(89)	180(81)	244(73)
	117(63)	181(55)	245(**47**) ------------(I)
54(45)	118(**37**)	182(**29**)	
55(**19**)	119(**11**)	183(**3**)	
	124(88)	188(80)	252(72)
	125(62)	189(54)	253(46)
62(44)	126(36)	190(28)	
63(18)	127(10)	**191**(**2**) -----------------------(0)	
	132(87)	196(79)	260(71)
69(69)	133(61)	**197**(**53**) -------------------------(VII)	
70(**43**)	134(35)	198(27)	
71(**17**)	135(9)	**199**(**1**)	
	140(86)	204(78)	268(70)
77(68)	141(60)	205(52) -------------------------(VI)	

78 (42)	142 (34)	206 (26)
79 (16)	143 (8)	207 (0)

Key to Table 5:

Z = Atomic number

$Z'h = Z + 9(Hmax)$; Hmax calculated as per equation 2, after Flint[3].

Bold numbers denote Z values through Z=57 which are prime numbers and Z'h values paired with these Z values which are also primes.

Figure 1.

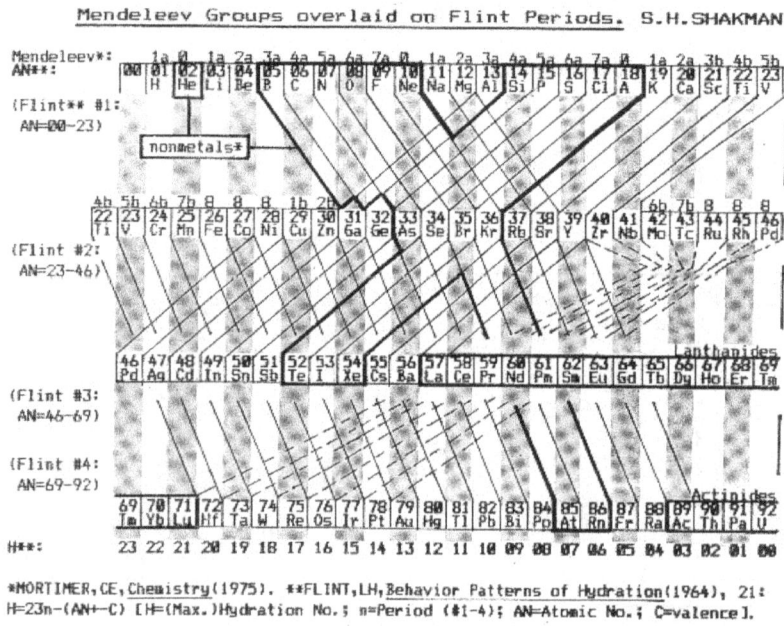

Mendeleev Groups overlaid on Flint Periods. S.H.SHAKMAN

*MORTIMER,CE,Chemistry(1975). **FLINT,LH,Behavior Patterns of Hydration(1964), 21:
H=23n-(AN+-C) [H=(Max.)Hydration No.; n=Period (#1-4); AN=Atomic No.; C=valence].

Key to Fig. 1:

Mendeleev* = *Mendeleev group*
AN = Atomic number
H** = Maximum hydration number per Flint (see Eq. 2)

References:

1. Maddox, J., Nature **359**, 669 (1992).

2. Maddox, J., Nature **347**, 13 (1990).

3. Flint, L. H., Behavior Patterns of Hydration (Institute for

   Advancement of Science and Culture, New Delhi, 1964), (a)

30ff, (b) 39ff, (c) 130-8, (d) 16, (e) 18, (f) 19, (g) 25,

(h) 20-22, (i) 80, 92, (j) 119.

4. Flint, L. H., J. Wash. Acad. of Sci. **22**, 97-119, 211-217 & 233-237 (1932).

5. Arrhenius, S., Theories of Chemistry, 91 (Longmans, Green and Co., London, 1907).

6. Robinson, F., in Encyclopedia Brit. 15th Ed. (1988), **18**, 268.

7. Weingärtner, H. & C. A. Chatzidmitriou-Dreismann, Nature **346**, 547-550 (1990).

8. Covington, A., Nature **351**, 534 (13 June 1991).

9. Amiss, E. S. and J. F. Hinton, Solvent Effects on Chemical Phenomena (Academic Press, New York and London, 1973), 52-54 (Table 3.1), 65-67 (Table 3.4), 109-110 (Table 3.15).

10. Graham, Thomas, Proc. Roy. Soc. **12**, 611-623 (1863), p. 614.

11. Flint, L. H., Hydration and Biology, 104ff (Institute for the Advancement of Science and Culture, New Delhi, 1968), (a) 104ff, (b) 45ff.

12. Flint, L. H., Dissenting Ape, 25-27 (Carlton, N. Y., 1973), (a) 53, (b) 25-27.

13. Flint, L. H., Plant Physiology **9**, 107-126 (1933).

14. Flint, L. F., Science **80**, 38-40 (1934).

15. Flint, L. F., and McAlister, E. D., Smithsonian Miscellaneous Collections **94**, No. 5, 1-11 (Publication 3334, 14 June 1935); and **96**, No. 2, 1-9 (Publication 3414, 16 June 1937).

16. Flint, L. and Moreland, C. F., Am. J. Botany **26**, 231-3 (1939).

17. CRC Handb. of Chem. and Phys., D167-8 (Chemical Rubber Co., Cleveland, Ohio, 1985-6).

18. Abegg and Bodländer, Zeit. f. Anorg. Chem. 454-499 (1899).

19. Maddox, J., Nature **354**, 263 (28 November 1991).

20. Fuller, R. B., Synergetics 2 (MacMillan, N. Y., 1979), 403-4.

21. Maddox, J., Nature **356**, 13 (5 March 1992).

22. Bredig, G., Zeitschr. Phys. Chem. **13**, 262 (1894).

23. Nernst, Walter, Theoretical Chem., 397 (Macmillan and Co.,

Ltd., London, 1916).

24. Senter, G., Trans. Faraday Soc. **3**, 146-152 (1907).

25. Bayliss, W. M., Principles of General Physiology, 177-8
(Longmans, Green and Co., London, 1915).

26. Perlzweig, W. A., Principles of the Theory, in Leonor
Michaelis, Hydrogen Ion Concentration, 125 (Baltimore,
Williams and Wilkins Co., 1926).

27. Pauling, L., Chemical Bond, 174 (Cornell U., Ithaca, 1967).

28. van't Hoff, J. H., Zeitschr. f. Phys. Chem. **i**, 481-508 (1887)
in Alembic Club #19 (Alembic Club, Edinburgh, 1929).

29. Kohlrausch, F., Gottingen Nachrichten, 213 (1876) in Goodwin,
Scientific Memoirs (MIT, Boston).

30. Einstein, A., Zeit. f. Elektrochem. **14**, 235-9 (1908), per
Theory of the Brownian Movement, 84-5 (Dover, 1956).

31. Bousfield, W. R., Proc. Royal Soc. **88A**, 147-169 (1912).

32. Encyclopedia Britannica, 15th Ed. (1991), Vol.1, p. 676.

33. Moseley, H. G. J., Phil. Mag., 703 (1914).

34. Bertsch, G. F. and S. McGrayne, in Encyclopedia Britannica,
15th Ed. (1991), Vol. 14, p. 330.

35. Bousfield, W. R., Phil. Trans. **206A**, 101-150 (1906), p.124.

36. Gluekauf, E., Faraday Soc., Transactions **51** 1241 (1955).

37. Ganong, W. F., Review of Med. Phys. 12 (Lange, Los Altos,1975).

38. Hildebrand, J. H., Reference Book of Inorg. Chem., Revised Ed.
37 (Macmillan, N.Y., 1940).

39. Chandrasekhar, J., D. C. Spellmeyer and W. L. Jorgensen, J. Am.   Chem. Soc. **106**, 903-910 (1984), (a)
p. 905, (b) p. 908.

40. Mason, B. J. in Kepler, J. (1611), A New Years' Gift or On The
Six-Cornered Snowflake 54 (Clarendon Press, Oxford, 1966).

41. Kepler, J. (1611), A New Years' Gift or On the Six-Cornered
Snowflake (Clarendon Press, Oxford, 1966), (a) 29, (b) 41,
(c) 23, (d) 45, (e) 43.

42. Pauling, L., Encyclopedia Britannica, 15th Ed. (1991), **15**, 938. 43. Eigen, M. and DeMaeyer, L., Proc. R.
Soc. **A247**, 505-533 (1958).

44. Amato, J., Science **255**, 255-6 (1992).

45. Shibata, K. and R. Matsumoto, <u>Nature</u> **353**, 633-5 (1991).

46. Watson, J. D. and F. H. C. Crick, <u>Nature</u> **171**, 737-8 (1953).

47. Fuller, R. B., <u>Synergetics</u> (MacMillan, N. Y., 1975), 133.

48. Max, N., <u>Nature</u> **355**, 115-6 (9 January 1992)

49. Einstein, A., (Dec.1954) <u>Meaning of Relativity</u>, 5th Ed.

    (Princeton U. Press, Princeton, 1956), 160-1.

Incorporates material from #S08279

Miscellaneous notes in article file
Checklist of subjects:
conductance
acidity-alkalinity index
relative hydration number
specific gravity
osmosis
gas solubility
elemental solid density
solute ionic size
crystal ionic size
energies of solution
snowflake structure
chlorophyll structure

- From Ziemilis, 19 Jan. 1993 from Nature:

"In reply please quote:
S08279A KZ/sjl

19 January 1993

Dr. S H Shakman
PO Box 382
Santa Monica, CA 90406-0382

Dear Dr Shakman,

Thank you for your letter of 24 December regarding your manuscript, "Conductance, hydration and periodicity". We have now had an opportunity to reconsider your paper in light of the changes you have made, but I am afraid that we do not feel able to alter our original decision regarding publication. I am sorry that we cannot respond more positively.

Yours sincerely,

 /initial LG, for/

Karl Ziemelis
Assistant Editor

- To Ziemilis, 4 Feb. 1993

> P.O. Box 382
> Santa Monica, CA, U.S.A. 90406-0382
> 4 February 1993

Karl Ziemelis, Assistant Editor
Nature
4 Little Essex Street
London WC2R 3LF

Reference: S08279A KZ/sjl

Dear Karl Ziemelis,

Thank you for your letter of 19 January 1993, which referred to your having reconsidered

"Conductance, hydration and periodicity" (#S08279), returned two copies to me, and indicated concurrence with Dr. David Lindley's 21 August 1992 communication to me (copy attached for your convenient reference). Please be advised that "Conductance ..." is the identical manuscript returned by Dr. Lindley on 21 August, that no changes had subsequently been made in it, and that my 24 December letter did not intend to seek reconsideration of it.

Also enclosed in your envelope were two copies of "Not just a flake model: hydration - conductance to quadruple helix", which was submitted on 24 December 1992, and to which Dr. Lindley's Aug. 21 letter does not apply. This later manuscript seeks particularly to respond to John Maddox's more recent and recurrent call for models which are both calculable and comprehensible [Nature **359**, 669 (22 October 1992), "The model for almost all seasons"; and Nature **347**, 13 (1990), "Virtue in imperfect models"]; however, your letter neither referred to "Not just a flake model..." or its return, nor offered an explanation as to how it may have failed to address the Maddox theme.

It would appear that my failure to designate dates of submission directly on my manuscripts may have resulted in your office having given attention to the wrong one, i.e. that returned by Dr. Lindley in August. I sincerely regret any inconvenience that may have been caused, but am hopeful this will not prevent my later submission from being considered on its own, relative to its intended scope as stated in its heading and introductory paragraphs.

Accordingly I respectfully request that "Not just a flake model: hydration - conductance to quadruple helix" be sent for review (five properly dated copies are enclosed for this purpose), or that you please be so kind as to identify any deficiencies which may prevent this so that an attempt might be made at correction.

Thank you for your kind attention.

> Sincerely yours,
> Stuart Hale Shakman

- From Nature, Postcard dated 9 Feb 1993, acknowledging receipt of "Not Just a Flake Theory" and assigning #SO2657

- To John Maddox, 13 October 1993

> P.O. Box 382
> Santa Monica, CA 90406-0382
> 13 October 1993

John Maddox, Editor
Nature
London

Dear Sir:

Re. handwaving/calculability of aqueous solution phenomena and snowflake formation [Maddox, J. Nature **364**, 483], your kind attention is referred to manuscript # **SO2657** (Nature submission), "Not just a flake model: hydration - conductance to quadruple [quad-looped] helix".

Could you also please advise status of review of # **SO2657**?

Thank you for your consideration and assistance.

> Sincerely yours,
> Stuart Hale Shakman

John Maddox, Editor, Nature **364**, 483 (5 August 1993),
"Paper crystals of molecular hydrogen and ice":

"It is not surprising that the simplest materials, bulk water and molecular hydrogen in particular, are among those whose bulk structures are the most perplexing: the mass of the hydrogen atom is only a third of that of the next heavier atom capable of forming molecules with ease. And even lithium can be a nuisance ... Low mass, of course implies ... mobility in the classical sense ...

"... Liquid water is the most common of all solvents, but its versatility is still better explained by hand-waving than by calculation. ...

...

Maddox refers to a study which identifies "a body-centered substructure in unit cell of hexagonal ice (the stable form at ordinary pressure, and from which snowflakes are formed) ... "

- To Judy, Nature, 12 Jan. 1994

> P.O. Box 382
> Santa Monica, CA 90406-0382
> (310)453-7707
> 12 January 1994

Nature
1234 National Press Building
Washington, D.C. 20045

Attention: Judy
Reference: Nature manuscript SO2657, "Not just a flake model ..."

Dear Judy,

As per my phone call of 28 December 1993 and followup of 11 January 1994, thank you for your help in attempting to trace my above-referenced manuscript. Enclosed for your convenient reference is a photocopy of the confirmation postcard dated 9 February 1993, which was the last communication received from Nature regarding this paper.

If, as you have suggested, a possible subsequent communication from Nature (London) to me regarding this paper may have been lost in the mail, could you please request that a duplicate be sent to me? Alternatively if the paper remains under consideration, please advise as to whether there is any action on my part that might facilitate this or to whom in London I might appropriately address a followup inquiry.

Thank you again for your kind assistance in this matter.

> Sincerely yours,
> Stuart Hale Shakman

enclosure

- From K. Ziemilis, 4 Feb. 1994 -- ???? missing; locate in file

---

- To John Maddox, Nature, 18 Feb. 1994

> P. O. Box 382
> Santa Monica, CA 90406-0382
> 18 February 1994

John Maddox, Editor
Nature
4 Little Essex Sreet
London WC2R 3LF

Subject: Calculability of aqueous solution phenomena through mathematical characterization of the hydrate theory of solutions

References:
  (a) My letter to you, 13 October 1993
    (b) Postcard from Nature, 9 February 1993
(copies enclosed for your convenient reference)

Dear Sir:

  Regarding my manuscript which discusses the above subject and was designated in above references as "SO2657": subsequent followup has disclosed that this number is not correct and that this manuscript is in your computer as "SO8279A".

  I regret any inconvenience that may have been caused by my citation of an incorrect number, and once again refer your kind attention to this manuscript. Or you can keep waving your hands.

> With kind regards,

Sincerely yours,
Stuart Hale Shakman

enclosures

Having heard nothing by October, the following was sent to John Maddox, the editor, who had suggested that our ability to calculate aqueous solution phenomena was less accurate than handwaving:

P.O. Box 382
Santa Monica, CA 90406-0382
13 October 1993

John Maddox, Editor
Nature
London

Dear Sir:

Re. handwaving/calculability of aqueous solution phenomena and snowflake formation [Maddox, J. Nature **364**, 483], your kind attention is referred to manuscript # **SO2657** (Nature submission), "Not just a flake model: hydration - conductance to quadruple [quad-looped] helix".

Could you also please advise status of review of # **SO2657**?

Thank you for your consideration and assistance.

Sincerely yours,
Stuart Hale Shakman

---

28 December 1993, having heard nothing, placed phonecall to London, with letter followup:

P.O. Box 382
Santa Monica, CA 90406-0382
(310)453-7707
12 January 1994

Nature
1234 National Press Building
Washington, D.C. 20045

Attention: Judy

Reference: <u>Nature</u> manuscript <u>SO2657</u>, "Not just a flake model ..."

Dear Judy,

   As per my phone call of 28 December 1993 and followup of 11 January 1994, thank you for your help in attempting to trace my above-referenced manuscript. Enclosed for your convenient reference is a photocopy of the confirmation postcard dated 9 February 1993, which was the last communication received from <u>Nature</u> regarding this paper.

   If, as you have suggested, a possible subsequent communication from <u>Nature</u> (London) to me regarding this paper may have been lost in the mail, could you please request that a duplicate be sent to me? Alternatively if the paper remains under consideration, please advise as to whether there is any action on my part that might facilitate this or to whom in London I might appropriately address a followup inquiry.

   Thank you again for your kind assistance in this matter.

Sincerely yours,
Stuart Hale Shakman

enclosure
_____

"In reply please quote:
SO8279A KZ/lw

4 February 1994

Dr S H Shakman
PO Box 382
Santa Monica, CA 90406-0382

Dear Dr Shakman

Thank you for your letter of 12 January, and my apologies for the length of time that it has taken us to get back to you. Unfortunately, there appears to have been a misunderstanding caused by an error in my letter of 19 January 1993. Unfortunately, when we received the revised and shortened version of your original submission, "Conductance, hydration and periodicity", the title of the paper was not changed on our database; as a result, the incorrect title was inserted in my previous letter. In other words, the paper, "Not just a flake model" hydration - - conductance to quadruple helix" was in fact considered on its own merits, but I regret that we judged it inappropriate for publication in <u>Nature</u>. I am sorry for any confusion that my previous leter may have caused in this regard.

I enclose with this letter the remaining copies of your manuscript, and I apologise for the confusion caused by our error.

Yours sincerely,

Karl Ziemelis
Assistant Editor

[copies were enclosed; top one had number "SO2657" erased but still visible]

P. O. Box 382
Santa Monica, CA 90406-0382
18 February 1994

John Maddox, Editor
Nature
4 Little Essex Sreet
London WC2R 3LF

Subject: Calculability of aqueous solution phenomena through mathematical characterization of the hydrate theory of solutions

References:
   (a) My letter to you, 13 October 1993
      (b) Postcard from Nature, 9 February 1993
 (copies enclosed for your convenient reference)

Dear Sir:

Regarding my manuscript which discusses the above subject and was designated in above references as "SO2657": subsequent followup has disclosed that this number is not correct and that this manuscript is in your computer as "SO8279A".

I regret any inconvenience that may have been caused by my citation of an incorrect number, and once again refer your kind attention to this manuscript. Or you can keep waving your hands.

With kind regards,

Sincerely yours,
Stuart Hale Shakman

enclosures

Correspondence with Prof. Howard Reiss, UCLA Chemistry Dept., 5 Jan -16 Feb 1996

5 January 1996 - hand-delivered:
  Shakman, S.H., <u>Nature</u> **338**, 456, "Heliocentric Tangents"
  Published abstracts (A E G K L M O P, GeoCities Athens 3361)

8 January 1996, letter to Dr. Reiss

11 January 1996, letter to Dr. Reiss correcting spelling of name.

30 January 1996, followup letter to Dr. Reiss

29 January 1996, letter from Dr. Reiss, crossed in transit.

16 February 1996, response to Dr. Reiss letter.

30 February 1996, conversation and notes thereon.

- - - - -

P.O. Box 382
Santa Monica, CA 90406-0382
8 January 1996
(310)453-7707

Dr. Howard Riess
UCLA
Chemistry Department

Dear Professor Riess:

  Thank you for your kind offer to read and provide written reaction to the materials I left with you on the subject of hydration. Your initial indication of objection to the system described therein, based on its apparent "heuristic" quality, was particularly appreciated insofar as this has already revealed a critical deficiency in those materials. In retrospect I can see how the apparent simplicity of the involved system might lead one to the assumption that it is based on some preconceived theoretical scheme; however, it must be emphasized that this is not the case.
  It is a bit ironic and even amusing that, in my effort to present a simplified picture and some of the potentially broad implications of Lewis Flint's work on hydration, I have omitted an essential discussion of how in the first instance this work was based on, and calculated directly from, measurements of equivalent conductivity for $K+$, $Na+$ and $Li+$ ions. The actual set of circumstances was marvelously unique, albeit somewhat involved, and I will try to describe it within a general historical context with which you are undoubtedly to some extent already familiar.

The essential components of this puzzle appear to be (1) the hydrate theory of solutions, an 1899 article by Abegg and Bodlander, and, according to Dr. Flint, the Stock Market Crash of 1929 which created the situation whereby he was exposed to Abegg and Bodlander; (2) the analogy between solute units and gases, as established through the works of Kohlrausch and van't Hoff, etc., which allowed Flint to attempt the use of Graham's law in conjunction with conductance and hydration; and (3) the fact that Flint had access and referred to a 1915 physiology text which contained essential data for his initial set of calculations, and not the 1924 edition of this same text which omitted some of this data.

As early as 1833, Sir Thomas Graham had spoken of "affinity for water" and a "theory of 23 atoms water" within the context of an experiment involving arseniate of soda.[1] Subsequently, the early development and advocacy of the so-called hydrate theory of solutions is closely associated historically with Dimitri Mendeleev (1887)[2]. Not long afterwards, in 1894, Bredig used the hydrate concept to explain why the chlorine ion is not more mobile than the heavier iodine ion, attributing this to a larger hydration shell around the Cl- ion that migrates with it[3]; this explanation has remained essentially unchanged up to modern times[4].

In 1899 Abegg (of "valence eight rule" fame) and Bodlander had sought to explain why mobility of the K+ ion is greater than that of the smaller and lighter Na+ ion, and the latter greater than that of the even smaller and lighter Li+ ion, with the suggestion that the hydration potential of these ions may vary inversely with their anhydrous weights[5]. This would allow for the hydrated Li+ ion to be the heaviest and least mobile of the three ions, and the much-less-hydrated K+ ion to be the lightest and most mobile. Over the next few decades, other investigators similarly invoked the concept of hydration in discussions of conductivities[6-9]; however, it was not until 1932 that a mathematical framework for Abegg and Bodlander's suggestion was advanced[10]. This occurred as an indirect consequence of the Stock Market Crash of 1929 and economies forced upon U.S. Government agencies: Lewis Flint was on loan from one U.S. Government office to another, where he was abstracting French and German articles on electrical measurements for Biological Abstracts and happened to come across the Abegg and Bodlander article.[11] At this point Flint became curious as to the possible use of Graham's law to delineate the relationship between conductance and hydration.[12]

In accord with Graham's law of diffusion (or effusion, of gases), the mobility of a gas varies inversely with the square root of its (density or) molecular weight[13], a relationship required by the more recent law of kinetic energy[14] (Graham himself credited Professor John Robison with having deduced this "pneumatic law" directly from the pre-Newtonian theorem of Torricelli on the velocity of efflux of fluids[13].)

As solute behavior had been shown to be analogous to gaseous[15], and conductivity was recognized as an index of mobility[16], Flint was able to view conductivity values as a measure (inverse-square-root) of relative hydrated weights of solute ions. In essence, Flint's work treated the Wheatstone bridge, which measures conductivities, as a "scale" for "weighing" solute ions.

In his initial work in 1932[10] Flint cited Nernst conductivity values for K+, Na+ and Li+ of 65.3, 44.4, and 35.5, resp., as listed W. Bayliss's 1915 physiology textbook[7] (this same Walter Nernst was specifically remembered by Einstein for his "witty" use of the Wheatstone Bridge[17]). On an adjoining page of the Bayliss text were listed W. Bousfield's proposed hydration numbers 4, 8 and 16, respectively, for these same ions[18].

While Flint did not specifically mention Bousfield, Flint's calculations were likely facilitated and to some extent even inspired by the presence of Bousfield's numbers in the 1915 Bayliss edition, whose date of publication corresponds with Flint's graduate studies at the U. of Vermont and Harvard; interestingly, specific reference to Bousfield's proposed hydration numbers was omitted from a subsequent (1924) edition of Bayliss's book. We may speculate that when Flint came across the Abegg and Bodlander article, his curiosity was to some extent sparked by recollection of Bousfield's proposed hydration numbers within the Bayliss text, which text he may have retained from the time of his graduate studies; used these numbers in preliminary calculations of relative mobility via Graham's law; and compared these calculated values with the Nernst data. The use of Bousfield's numbers would have given results that were encouraging but not precise; however, a minor adjustment from Bousfield's numbers to an alternate set of numbers, 4, 12, and 20 respectively, yields an optimum fit with the Nernst data, via Graham's law.

Flint found that presumed hydration numbers of 4, 12, and 20, each multiplied by 18 (for the weight of each water unit) and added to conventionally- appraised atomic weight values of (approx.) 39 for potassium (K), 23 for sodium (Na) and 7 for lithium (Li), resp., yielded totals (111, 239 and 367) whose relative inverse-square-roots (.0949, .0647 and .0522) correlated remarkably well with the Nernst data - almost unbelievably well.

Flint then noted that each of these initially-hypothesized hydration numbers (4, 12, and 20), when added to associated atomic numbers (19, 11 and 3 for K, Na and Li, resp.) equalled 23.

This was the basis for Flint's initial conception that, for the lighter ions, an inverse, reciprocal and integral relationship between atomic number and hydration number appeared to exist; and for his subsequent proposed refinements, extension through the 92 naturally-occurring elements, and interpretation that these hydrational patterns may also be periodic.

This is where it all began. Flint's subsequent work over a period of four decades sought to expand on this original observation concerning the K+, Na+ and Li+ ions, and he was the first to assert that no aspect of his work was beyond improvement or correction. However, the fact remains that in the first instance this work is based not on theory but on measurement. To the extent that it apparently has "heuristic" value, this is not something that was originated by Flint, but rather based on that which he observed and sought to further describe. That the resulting methodology allows for calculation of various phenomena from first principles of atomic number and valence is certainly understandable grounds for cynicism, but not in-and-of-itself proper grounds for rejection. Rather, as the methodology is algebraic and hence very computer-compatible, it is readily accessible to exhaustive examination and challenge; assuming such further examination does not result in its refutation, it would appear to offer particular advantages to whoever might be in a position to begin to explore it more fully at this early stage of development.

I am constantly reminded and encouraged by the fact that 18 centuries had passed before Copernicus revived Aristarchus's heliocentric hypothesis; in contrast the mere 6 decades that have passed since Flint's original work seem insignificant and, likewise, would not seem to constitute sufficient grounds to reject out of hand the suggestion that Flint's work may merit at least further consideration at this time.

I hope that this somewhat windy historical perspective (1) might demonstrate that I have not uncritically accepted Flint's admittedly controversial work, but rather have examined its historical foundations (as well as into its internal consistency as might be inferred from the materials I gave you on Friday last); and (2) most importantly that it might encourage you to begin to view this work as a potential tool which may complement and augment ongoing studies of solute phenomena. I would welcome the opportunity to share with you, your colleagues and students those materials I have been able to access relating to Dr. Flint's work and the products of my own efforts over the past decade to further validate and extend it.

Again, I thank you for your kind consideration and time.

Sincerely yours,
Stuart Hale Shakman

1. Graham, Sir Thomas, FRS, Researches on the Arseniates, Phosphates, etc. (1833), p. 15, 45.

2. Arrhenius, S., Theories of Chemistry, 91 (Longmans, Green and Co., London, 1907).

3. Bredig, G., Zeitschr. Phys. Chem. **13**, 262 (1894)

4. Robinson, F., in Encyclopedia Brit. 15th Ed. (1988), **18**, 268.

5. Abegg and Bodlander, Zeit. f. Anorg. Chem. 454-499 (1899).

6. Senter, G., Trans. Faraday Soc. **3**, 146-152 (1907).

7. Bayliss, W. M., Principles of General Physiology, 177-8 (Longmans, Green and Co., London, 1915).

8. Nernst, Walter, Theoretical Chemistry, 397 (Macmillan and Co., Ltd., London, 1916).

9. Perlzweig, W. A., Principles of the Theory, in Leonor Michaelis, Hydrogen Ion Concentration, 125 (Baltimore, Williams and Wilkins Co., 1926).

10. Flint, L. H., J. Wash. Acad. of Sci. **22**, 97-119, 211-217 & 233-237 (1932).

11. Flint, L. H., Dissenting Ape, 8 (Carlton, N. Y., 1973).

12. Flint, L. H., Behavior Patterns of Hydration (Institute for Advancement of Science and Culture, New Delhi, 1964).

13. Graham, Thomas, Proc. Roy. Soc. **12**, 611-623 (1863), p. 614.

14. Pauling, L., Chemical Bond, 174 (Cornell U. Press, Ithaca, N.Y., 1967).

15. van't Hoff, J. H., Zeitschr. f. Phys. Chem. **i**, 481-508 (1887) in Alembic Club #19 (Alembic Club, Edinburgh, 1929).

16. Kohlrausch, F., Gottingen Nachrichten, 213 (1876) in Goodwin, Scientific Memoirs (MIT, Boston).

17. Einstein, A. (1942), in Out of my Later Years, 233 (Philosophical Library, New York, 1950).

18. Bousfield, W. R., Proc. Royal Soc. **88A**, 147-169 (1912).

P.O. Box 382
Santa Monica, CA 90406-0382
30 January 1996
(310)453-7707

Dr. Howard Reiss
UCLA
Chemistry Department

Dear Dr. Reiss:

With reference to my letter of 8 January 1996 and materials left with you on 5 January 1966, this letter is intended to (a) illustrate how work related to and prior to that which enabled Flint's initial calculations in 1932, as discussed in the 8 Jan. letter, apparently continues to influence contemporary work in chemistry on related subjects; (b) provide some information on my background, and relate how I came to study the subject of hydration; and (c) clarify my goal in having approached your department and suggest a possible course of action.

(a) D. R. Rosseinsky's recent discussion of electrostatic continuum models (J. Am Chem. Soc. **1994**, 116, 1063-1066) cites two of his prior works (J. Chem. Soc. (A: Inorg., Phys., Theor.) **1971**, 608-610; and Electrochim. Acta **1971**, 16, 19-22), the second of which is readily available in the UCLA Chemistry Library. Therein Rosseinsky cites the 1907 work of W. R. Bousfield and T. M. Lowry, Trans. Faraday Soc. **3**, 123) along with a 1965 Rosseinsky paper which also prominently cites the 1907 Bousfield and Lowry paper (D. R. Rosseinsky, Chem. Rev. **65**, 467).

As you may please note, my letter to you of 8 January describes how Bousfield's later work on the hydration number (published in 1912 and repeated in Bayliss's 1915 textbook) apparently facilitated or even inspired Flint's 1932 work, in conjunction with some other "serendipitous" events.

(b) After graduating from Northwestern U. in 1964, I attended Georgetown U. graduate school in a Ph.D. program in History of Political Theory, where I was inducted into Pi Sigma Alpha, the national political science honorary society. My graduate studies were interrupted due to travel requirements of my job as program officer for the Vietnam health program with the U.S. Department of State's Agency for International Development (A.I.D.). While at A.I.D. I was awarded a Meritorious Service Increase.

My analysis of Flint's methodology has drawn upon skills developed at A.I.D. and in other professional positions which followed, which work has required me to be both thorough and critical. Perhaps as significantly, the A.I.D. experience educated me as to the validity and importance of the role of the generalist in analyzing technical programs, and gave me the confidence to undertake assessment of Flint's work. For your reference, I am enclosing a copy of my personal resume.

My involvement with Flint was itself quite "serendipitous", evolving from an investigation into a connection between sea-floor vents and the origin of petroleum, and the related conceptual possibility that gases from these vents might, through a process of forced hydration, somehow relate to the origin of life. A search at the Library of Congress for a connection between "hydration" and "biology" turned up Flint's book <u>Hydration and Biology</u> and then his preceding volume <u>Behavior Patterns of Hydration</u>, where I found his statement:

"At the time of the discovery in 1932, I had attained the age of 39, an age held and maintained by no less an authority than Jack Benny to represent the very peak of perfection in a human male. Under the circumstances there was engendered a measure of personal pride and enthusiasm, and basic mathematical data for weight, hydration and mobility covering the four periods were prepared. ..." .

At the time of my first encounter with Flint's work in 1982, I was 39, and I will confess that his genuine manner as reflected in this statement was a factor in my decision to attempt a fair assessment of his work.

Following a preliminary determination that the fundamental work appeared to be mathematically sound, I contacted L.S.U. where Flint had taught botany, and was informed that he had died in 1973 and that further work on hydration was not being conducted there. This also led to contact with Flint's family and access to additional biographical and bibliographical materials. Subsequently I have: searched for anything that might repudiate, refer to, build on or transcend Flint's work, or otherwise render it obsolete - to no avail; reviewed his work more thoroughly and sought to extend it; attempted to supplement deficiencies in my background knowledge as required, both technical and historical; sought to extend Flint's methodology to phenomena not addressed by him; and continued to monitor the general scientific literature on related topics. The materials given you are the "tip of an iceberg" of related studies which I have attempted to chronicle along the way.

  (c) It was not my intention in my 8 January letter to present myself as a "Mr. Know-It-All" the subject of hydration (after Flint), but rather as an ardent student on the subject over a period of several years. I am interested in further developing these studies within an academic framework that involves properly rigorous scrutiny and review; interacting with others who are investigating related questions and incorporating considerations of how Flint's methodology may (or may not) be complementary; and structuring these studies within proper dissertation and teaching formats. It is my hope to open a dialogue with you on whether and how this might be attempted at UCLA, under the primary guidance of your department.

  Perhaps this might be approached from a "history of science" perspective, with emphasis on the hydrate theory of solutions and Flint's proposed quantification. Such an approach might both (1) be facilitated by the circumstance of my previous graduate work at Georgetown in "history of political science" (this of course wishful thinking on my part), and (2) appropriately characterize my "history of the hydrate theory" studies as a supplement rather than a challenge to contemporary work on solutes in aqueous solution.

  As I mentioned when we briefly spoke on 5 January, I have spoken at length with Dr. John Prausnitz at U.C. Berkeley, who has suggested that I target the <u>Journal of Chemical Education</u> for an historically-based article on hydration, and has kindly offered to review my draft. Work on such an article has begun, with the benefit of the perspective gained from my initial contact with you as discussed in my 8 January letter; however, it would appear highly desirable to also continue to seek to integrate my studies within the broader context of an established program of graduate study, as discussed above. I am hopeful that this letter allows the materials I have given you to be regarded as background materials which do not necessarily require a formal response, although I assure you that your response in any form is eagerly awaited and will be greatly appreciated.

As I rapidly approach my 53rd birthday, a milestone that was the last celebrated by my late father, I admit to experiencing a sense of urgency in this endeavor. Nonetheless, I hope you will consider this letter as primarily motivated by a desire to clarify my goal, which clarification will hopefully facilitate your response.

Thank you again for your time and consideration.

> With kind regards,
>   Sincerely yours,
>   Stuart Hale Shakman

encl.

- - - - -

Howard Reiss
Professor of Chemistry
UCLA
January 29, 1996

Mr. Stuart Hale Shakan
P.O. Box 382
Santa Monica, CA.

Dear Mr. Shakman:

Please forgive this late response to the material that you left with me. As you can imagine, I have been extremely busy with my various commitments and could only give consideration to your writings and ideas on a part time basis

At the very outset, I have to tell you that I think it would be extremely difficult to find somebody who would be willing to collaborate with you in a research or teaching project embodying your idea. At the same time I want to tell you (without offering an artificial platitude) that I am quite impressed by the tenacity and enthusiasm with which you have embraced your concept and the degree of understanding that you have in spite of the fact that you have no formal training in chemistry and especially physical chemistry.

Your work (and Lewis Flint's work) can in no way be considered as a "crank" investigation of solution theory. It is, however, terribly oversimplified in view of our modern understanding of these phenomena. More than anything else, the scientific method depends on "induction" in which the essences of many phenomena are distilled into a few positive statements that cannot be ultimately proved absolutely, but which represent a set of axioms or "laws" that constitute a "description" of how we think things are. From these axioms we "deduce" theorems that predict the result of a measurement. The more accurate the predictions, the more confidence do we have in the underlying axioms (description).

Nowadays we have several disciplines, based on axioms, that have proved to be extremely accurate in providing predicitons, usually on the basis of molecular theory. These are subjects like quantum theory, statistical mechanics, and thermodynamics.

The correlations that you and Flint have found are just the sorts of things that are involved in the induction process that led to the axioms that underpin the above mentioned disciplines. For example, using these disciplines, we can compute, quite accurately what the boiling or freezing point of a substance will be, or what its molecular structure will look like. We can predict the conductivity, viscosity, or diffusivity of a species in an electrolyte solution and, yes, even its state of hydration (solvation) as well as the structure of that hydrate! In this prespective Graham's law of diffusion is somehat archaic (although it certainly applies to ideal gases). Indeed, the transport properties of molecules have been the subject of intense investigation even in the very complicated situations found with "complex fluids", e.g. polymers, microemulsions, liquid crystals, etc.

To form an idea of what I am saying you should look at any one of many books on physical chemistry, statistical mechanics or quantum mechanics. You might look, for example, at the following books:

1. "Theory of Simple Liquids" by J.P. Hansen and I.R. McDonald, Academic Press, 1986.
2. "Equilibrium Statistical Physics" by M. Plischke and B. Bergersen, 2nd edition, World Scientific, 1944.
3. "Statistical Physics of Macromolecules" by A.Y. Grosberg and A.R. Khokhlov, AIP Press, 1994.
4. "Intermolecular and Surface Forces" by J. Israelachvili, 2nd edition, Academic Press, 1991.
5. "Physical Chemistry" by J.H. Noggle, 3rd edition, Harper Collins, 1996.

Perhaps it would be best for you to start with the last two of these books, so that you can get a quick idea of what has been accomplished.

For you to make progress in your areas of interest it would be necessary, I believe, for you to acquire some formal training in the fields on which physical chemistry and chemical physics are based. How you will do this is of course a problem, but it is the best advise I can give you.

I am sorry that I cannot be more encouraging.

Yours Sincerely,
/s/
Howard Reiss, Professor of Chemistry

HR:pr

- - - - -

P.O. Box 382
Santa Monica, CA 90406-0382
16 February 1996
(310)453-7707

Professor Howard Reiss
UCLA
Chemistry Department

Dear Professor Reiss:

Thank you for your kind and thoughtful letter of 29 January 1996, which I received on 2 Feb. and which crossed in transit with my 30 January follow-up letter; and thank you also for your referral to contemporary books that relate to your work.

Your letter has highlighted the need to clarify at the outset that Dr. Flint's (and my associated) work on hydration, (1) as relates to theoretical considerations, is not nearly so ambitious as would be an investigation of solution theory, but rather involves a far more limited and specific investigation, within the context of the venerable "hydrate theory of solutions", of a mathematical framework that resulted from the use of conductance as a precise mathematical index of hydration; and (2) as such, is deductive in origin and rests on an empirical, not a theoretical, foundation.

For my part, after reading your letter and looking at those books you cited that were readily accessible to me, I am further encouraged that these works on hydration with which I am associated may usefully complement the work of the Chemistry department and others concerned specifically with the understanding of aqueous solutions. (How this may or may not relate to your work towards an understanding of solutions or liquids in general is clearly beyond my scope.) I will herein attempt to explain this to the best of my ability, with particular reference to your letter of 29 January and the materials cited therein that I was able to look at, i.e.: <u>Intermolecular and Surface Forces</u> by J. Israelachvili, 1985; <u>Physical Chemistry</u> by J.H. Noggle, 1996; <u>Equilibrium Statistical Physics</u> by Plischke and Bergersen, 1989; and <u>Theory of Simple Liquids</u> by Hansen and McDonald, 1976.

First, a word on how I came to knock on your particular door. On Friday, 5 January 1996, I went upstairs to the Physical Chemistry office area, specifically asked the two professors I encountered there if they or anyone else in the department might be familiar with the hydrate theory of solutions - an old theory that was prominent around the last turn of centuries - and was referred to you.

Beyond having originally evolved within the tradition of the hydrate theory of solutions, Dr. Flint's work and the results of his several years of investigations were regarded by him to particularly relate to processes generally considered within the domains of biophysics and biochemistry. But because the origin of his work rested on interpretation of conductivity of ions in aqueous solution, the domain of physical chemistry seems a logical starting place for consideration of it. Moreover, because the system of Flint had at first seemed to me to be far too "pretty" to be other than contrived - "overly simplified" if you will - I have sought to critically examine its foundations, and thus have been particularly exposed to historical and other considerations which appear to relate to the field of physical chemistry.

By its very nature, the hydrate theory of solutions is restricted to aqueous solutions, is inherently simple, and is readily visualized as inherently algebraic; it involves the concept of association of a specific number of identifiable whole water units to ions in solution. The hydrate theory was a field of intense investigation at the end of the last century and during the early years of this century, involving the likes of Ostwald, Kohlrausch, Nernst, van't Hoff, Arrhenius and many others, including some particularly exhaustive works by Harry Jones* of the Carnegie Institution.

Dr. Lewis Flint was, if you will, "heir" to this legacy, or at least intensively exposed to it during his formative years. With this background, it is understandable how Dr. Flint's reaction to the set of serendipitous circumstances that fell into his lap in 1932, as described in my letter of 8 January, was quite natural, even instinctive. He was not investigating solution theory, as such, but rather was simply using the relative mobilities of solute ions as an index of their relative hydrated weights - which calculation disclosed a simple relationship between atomic numbers and hydration numbers.

In contrast to much more complex and diverse modern attempts to understand phenomena involving solutes in aqueous solution, Dr. Flint's initial finding and the algebraic framework that seemed to follow from it might easily at first glance be considered overly simple; however:

(1) It is noted that the various approaches discussed in the books to which your have referred involve varying degrees of complexity; some are more complex than others. Just as each must stand or fall on its own merits, one would hope that Dr. Flint's (and my associated) work would not be rejected out of hand just because it may be less complex.

(2) It also appears that these various modern approaches or components of them are not always considered mutually exclusive, but rather are often combined in some fashion in an effort to explain phenomena under study, and modified in the process as seems warranted. Similarly, Dr. Flint's offerings in this area need not be viewed as an "either or" proposition, but rather as a tool which may usefully augment, and/or be usefully augmented by, these other approaches. Within the context of conventional chemistry, Dr. Flint regarded his work as "interpretive and supplementary".

(3) Totally aside from consideration of any merits or failings of Dr. Flint's or my work, I had been of the impression that actual understanding of these phenomena, even in modern times and with sophisticated modern methods, is far from complete. For example:

J. P. Hansen and I. R. MacDonald, Theory of Simple Liquids 1976, refer (p. 62) to A. Rahman and F. H. Stillinger 1971 [J. Chem. Phys. **55** 1971, p. 3336] on the subject of water, who in turn note that a "combination of complications renders impractical a large part of conventional liquid state theory for studying water."

In a very recent article, J. Israelachvili and H. Wennerström, "Role of hydration and water structure in biological and colloidal interactions", Nature **379**, 18 January 1996, 219-225, take issue with the "conventional explanation" that "hydrophilic surfaces and macromolecules remain well separated in water" due to a "repulsive hydration force owing to structuring of water molecules at the surfaces". The authors suggest an "alternative interpretation" whereby the "repulsions have a totally different origin."

Honig, Barry and Anthony Nicholls, "Classical Electrostatics in Biology and Chemistry", Science **268**, 26 May 1995, 1144-1149, note that "much remains to be done before [realizing the goal of] a complete and accurate method to describe the properties of molecules in aqueous solution".

Wheatley, DN, Nature **366** (25 November 1993), 308, "Water in life", states "as yet no calculable theory exists with which to begin to consider water's interactions with other molecules."

Maddox, John, Nature **364**, 483 (5 August 1993): "... Liquid water is the most common of all solvents, but its versatility is still better explained by hand-waving than by calculation."

Hammel, H.T.,J. Phys Chem. **1994**, *98*, 4196-4204, states "Two theories, one by G. Hulett in 1903 and one by G.N. Lewis in 1908, describe how solutes alter water in an aqueous solution. ... Neither Hulett's nor Lewis's theory incorporates a kinetic theory to account for the altered state of the solvent in a solution. ..." Hammel asserts " ... "Hulett's theory should supplant Lewis's theory as the preferred account of the altered state of the solvent in a solution."

Covington, A., Nature **351** (13 June 1991), 534, reviewing Thermodynamics of Solvation ..., 1991, by G.A. Krestov, notes that Krestov points out: "that the essential feature of solvation is a pictorial one of bonding of the solvent in the immediate vicinity of an ion forming a solvation shell. Problems arise in trying to determine the thickness of the shell and how many molecules it contains - the solvation number."

M. Berkowitz and W. Wan, Journal of Chemical Physics (1 January 1987. p. 377), "The limiting ionic conductivity of Na+ and Cl- ions in aqueous solutions: Molecular dynamics simulation", begin with "The mechanism of ionic transport is a classical problem in physical chemistry". The authors open their discussion of molecular theory with "To describe the limiting ionic mobility on a molecular level is a very challenging task. It is obvious that first theories will have to include simplifying assumptions ..."; and conclude their introductory abstract with "The major assumptions of molecular theory of the limiting ionic mobility were tested and not confirmed by the simulation."

- - - - -

   It is further noted that the potential role or even existence of hydrated ions does not uniformly enter into considerations of some phenomena. For example, two recent authoritative articles assert that the greater permeability of cell membranes to the potassium ion vs. the sodium ion involves considerations other than size, since the sodium ion has a smaller ionic radius [Jan, L.Y. and Jan Y.N., Nature **371** (1994), 119-122; and Kumpf, R. A. and D. A. Dougherty, Science **261** (1993) 1708-1710]; whereas Ganong's physiology textbook offers an explanation of permeability involving the consideration that these ions may be hydrated and that the hydrated sodium ion is actually larger than the hydrated potassium ion [W.F.Ganong, Review of Medical Physiology, 12th Ed., Lange Medical Pub., 1985, p. 21].
   (Without specific reference to permeability of membranes, Israelachvili 1985, p. 43, offers support to the Ganong position with the statement, "Because smaller ions are more hydrated they tend to have larger hydrated radii than larger ions.")
*****

Whereas some contemporary articles reject the concept that ion selectivity, through membranes, pores or so-called channels, is due to ionic size, others allow for this to be the case due to hydration; ions which are smaller in an anhydrous state may actually be larger when hydrated.

A. Two recent articles which reject the concept of ion selectivity based on size:

1. Jan, L.Y. and Jan Y.N., <u>Nature</u> **371** (8 Sept. 1994), 119-122, "Potassium channels and their evolving gates", p. 119,

"All potassium channels are at least 100 times more permeable to potassium ions than sodium ions ... .Since sodium has a smaller ionic radius than potassium, it is unlikely that selectivity is achieved by physical occlusion.

2. Kumpf, R. A. and D. A. Dougherty, <u>Science</u> **261** (24 September 1993, 1708-1710, "A Mechanism for Ion selectivity in Potassium Channels: Computational Studies of Cation-$_\pi$ Interactions"

 Kumph and Dougherty state "An important feature of the shaker channel and related structures is their considerable ion selectivity; K+ is preferrred over Na+ by a large margin and over larger ions such as Rb+ to a smaller but still significant extent. A major challenge in the ion channel field is the identification of the molecular interactions responsible for this selectivity."

"... Streaming potential measurements suggest that K+ is substantially desolvated as it traverses the membrane."

"Table 1. Gas-phase binding data" lists the following:

Ion	Ionic radius (A)	Aqueous solvation energy (kcal/mol)	M+-Cl- binding energy (kcal/mol)
Li+	0.60	122	153
Na+	0.95	98	134
K+	1.33	81	120
Rb+	1.48	75	115

Sidenote: we find the aqueous salvation energy data in Science Mag along with volumes; re the salvation energy data, it might be correlated with Flint's methodology thusly:

Speculative calculations - Aq. Solvat. Energy (kcal/mol) (data from <u>Science</u> Magazine)

	Ob	Assumption	Calc Error (ASE~=Sq/Rt.Z)	\\Vh [not related to set at left?]
Li+	122	Hmax	.01	171.1
Na+	98	"	.00 [base]	100.2
K+	81	"(2 K+)	.11	33.0
Rb+	75	0 (2Rb+)	.08	81.8
Cs+				

We note that Kumph and Dougherty provide only data pertaining to anhydrous ions, ...

 Whereas Kumph and Dougherty have sought to explain ion selectivity in so called potassium channels, in concert with theories of ion permeability which invoke a balance of "binding and dislocation effects [forces]", Ganong has suggested that differences in permeability of membranes may be explained on the basis of ion size, noting in particular that "the hydrated sodium ion, i.e., Na+ with its full complement of water, is larger than the hydrated potassium ion."

In this connection, conductance of ions has for 100 years been accepted as an index of size, with respective increasing conductivities of Li+, Na+ and K+ taken to indicate decreasing hydrated size.

In concert with Ganong's discussion of the possible relation between pore size and ion selectivity, preference for K+ over Na+, Li+ and Rb+ may be atttributed to respective hydrated size, as listed in ___ and as may be calculated in concert with L.H. Flint's proposed description of hydrational potentiality ....

B. In contrast, the possibility that the size of hydrated ions is the cause of such differentiation is discussed by Ganong and within a recent edition of the Encyclopedia Britannica:

W.F.GANONG, Rev.of Med.Physiology, 12th Edition (Los Altos, CA, Lange Medical Publications, 1985),p. 21]:   "Particle size ... affects the movement of ions across cell membranes, and it should be noted that the ions in the body are hydrated. Thus, although the atomic weight of potassium)39) is greater than the atomic weight of sodium (23), the hydrated sodium ion -- i.e., Na+ with its full complement of water -- is larger than the hydrated potassium ion.   However, it is clear that ions cross membranes via ion **channels** rather than simple pores. ..."

[W.F.GANONG, Rev.of Med.Physiology(1975),p. 12]:
 "The reason for the differences in the permeability of cell membranes to various small ions is unknown. It is tempting to speculate that the membranes contain pores and that the differences in permeability can be explained on the basis of ion size. ...
 "The ions in the body are hydrated, and although the atomic weight of potassium (39) is greater than that of sodium (23) the hydrated sodium ion, i.e., Na+ with its full complement of water, is larger than the hydrated potassium ion."

ENCYCLOPEDIA BRITANNICA : The first thing noticeable in a table of ionic mobilities in aqueous solution is that ions differing considerably in mass and size have much the same mobility. Thus, for example, the ions Cl-, Br-, and I- have almost identical mobilities although it might be expected that the heavy and bulky I- ion would have a much lower mobility than the lighter and smaller Cl- ion. Even more surprisingly, the small and light lithium (Li+) ion has only about half the mobility of the heavy cesium (Cs+ ion. This is generally attributed to the fact that ions collect a group of water molecules around themselves, which move as a unit with the ion." ...
 "... it is well established that the purple colour of potassium chromium alum crystals is caused by the influence on the chromium (Cr3+) ions of a coordinate octahedron of molecules of water of crystallization. Since this colour persists in aqueous solution, the coordinated octahedron must be retained in solution. Further, if the solution is heated (and the octahedron destroyed), the ions take up a new configuration that gives a green solution. Crystals of chrome alum cannot be obtained from this green solution."

*****

With reference to your statement that "we can predict the conductivity ... of a species in an electrolyte solution, and ... its state of hydration (solvation) as well as the structure of that hydrate", and with reference to the four books you cited, I note that none of these makes specific mention of the hydrate theory of solutions. However, discussions of hydrates in Israelachvili 1985, in your letter, and elsewhere in the literature, indicates the continued viability of the concept of hydrates in aqueous solution and thus the lasting legacy of the hydrate theory of solutions. Therefore a prospective clarification of hydration numbers existing under conditions incident to the measurement of ionic conductance would appear to properly fall within the scope of your department's area of interest.

On p. 43, Israelachvili 1985 states "Hydration numbers and radii can be deduced from measurements of the viscosity, diffusion, compressibility, conductivity, and various thermodynamic and spectroscopic properties of electrolytic solutions, the results rarely agreeing with one another (Amiss, 1975; Saluja, 1976)". Flint's initial work, which was the result of simple deduction (direct calculation) of hydration numbers from conductivity, certainly seems to fit within this framework.

As evident in your 29 January letter, my 8 January letter had apparently failed to adequately illustrate that Flint's initial finding was deductive. In the first instance, Flint treated relative mobility as an index (inverse-square-root) of relative hydrated weight; which relative hydrated weight would be comprised of components of atomic weight and weight of water-of-hydration: which "weight of water-of-hydration" components divided by 18 yielded the initial set of hydration numbers recorded by Flint. He certainly enjoyed some lucky circumstances which facilitated his calculation, as discussed in my letter of 8 January, but nonetheless the calculation itself yielded his result and disclosed the pattern which he reported and on which his subsequent work was based. Perhaps Equation 1 below will help illustrate this, wherein: "X", "Y" and "Z" are the sought-after hydration numbers (minimum whole numbers) for Li+, Na+ and K+; the number "18" represents the weight of water, against which "X", "Y", and "Z" are multiplied; 7, 23 and 39, resp., are atomic weights for the three ions; and 35.5, 44.4 and 65.3 are their limiting ionic conductivities per Nernst, as reported by Bayliss 1915.

EQUATION 1:

(1) $35.5: 1/\sqrt{(18X +7)} :: 44.4: 1/\sqrt{(18Y +23)} :: 65.3: 1/\sqrt{(18Z +39)}$

The results of Dr. Flint's calculations were X=20, Y=12 and Z=4, resp., which numbers when added to corresponding atomic numbers happened to each total 23. As noted in my 8 January letter, the fact that these totals happened to be 23 and the basis for simplicity inherent in the resulting system described by Dr. Flint were not of his invention, but rather the inescapable result of this calculation from mobilities. Dr. Flint was very surprised by this result, but, as he could not ignore it, he chose to explore it - and continued to do so over the next four decades.

As indicated in my 8 January letter and the other materials given you on 5 January, this calculation incorporated the principle formerly known by the "archaic" term "Graham's law" (the term in use in my high-school chemistry courses - has it really been that long?), which in any case describes a relationship that is required by and fully consistent with the law of kinetic energy and molecular theory. Your letter of 29 January enthusiastically recognizes the applicability of the principle in the case of ideal gases but not within the perspective of solutes in aqueous solution. This is understandable insofar as Dr. Flint was apparently the first and still the only one to so apply it, this on the strength of the principles of Kohlrausch and van't Hoff, etc., regarding the independent migration of ions in aqueous solution and establishing that solutes (presumably including such independently-migrating ions) in aqueous solution and gases in a gaseous environment behave in analogous fashion. The apparent indication that you are not familiar with the application in modern times of the "Grahams law" relationship in conjunction with aqueous solutions is in essence a confirmation that Flint was correct when he related that he was unaware of anyone having previously used it in this perspective, and also a confirmation of the thoroughness of my own continuing investigation indicating that no one independently of Flint has so used it since. This in-and-of itself does not seem to constitute a reason why it should not be so used.

On the contrary, it appears that a recent precedent may be found in Hammel's above-cited recent article [Hammel, H.T., J. Phys Chem. **1994**, *98*, 4196-4204], which concludes:

"The connection between the thermal motion of solute molecules bounded by a solution that contains them and the thermal motion of gas molecules contained in a vessel is not accidental. McMillan and Mayer [W.G. McMillan and J.E. Mayer J. Chem. Phys. **1945**, 13, 276] applied the method of the grand-canonical ensemble to multicomponent systems and obtained equations for the calculation of thermodynamic properties and distribution functions of such systems. The theory was applied to liquid solutions for which the osmotic pressure was shown to play a role analogous to that of the total pressure of a gaseous system."

Thus McMillan and Mayer's work might be considered as further validation of van't Hoff's demonstration of the analogy between solute and gaseous phenomena, which was also derived through the study of osmotic pressure and gaseous pressure (for which work van't Hoff was awarded the first Nobel prize in chemistry in 1901); and Hammel's work comprises a contemporary example of consideration of the potential utility of application of a principle that applies to the gaseous state within the perspective of solutes in aqueous solution. This would seem to particularly hold for so basic a principle as one that is required by the law of kinetic energy, e.g., the so-called "Graham's law" - which law enabled Dr. Flint to directly convert relative ionic mobilities into relative hydrated ionic weights (including calculable numbers of water units), and thus rests at the very foundation of the system described by him. In contrast, the proposed system described and advocated by Hammel is based on Hulett's 1903 thought experiment, which is not intended as an insult to the works of Hammel or Hulett, but rather as an argument for at least some measure of consideration of Flint's work.

It is further noted that McMillan and Mayer's validation of the analogy between solute and gaseous phenomena had involved use of the "grand-canonical ensemble", which Plischke and Bergersen 1989 discuss within their self-described "review of the basics of statistical mechanics". Thus it would appear that Dr. Flint's application of the "Graham's law" principle to solute ions in aqueous solution might be viewed as consistent with and validated by statistical mechanics.

The above reference to Hulett's thought experiment suggests that it may be useful to briefly discuss the visualizable nature of the operative mechanism of hydration described by Dr. Flint. While his specific methodology concerning hydration numbers was a result of the above-described direct calculation from conductivities, the general mechanism described by Flint seems to resemble one described recently by Israelachvili and Wennerström 1996 [Nature **379**, 18 January 1996, 219-225], p. 220, in the case of strongly hydrophilic molecules or surface groups. As described by Dr. Flint, and in these cases by Israelachvili and Wennerström, the dissociation of the solute occurs because the attraction between water and solute is stronger than the former solute-solute attraction. Subsequently in Flint's work, no additional repulsive forces (such as those discussed by Israelachvili and Wennerström 1996) were postulated or even discussed; rather, the solute units, whether strongly associated to a discrete number of water units (hydrated) or not, were assumed to have been liberated from any former attraction to like solute units by virtue of the stronger attraction for the solvent, and were presumed to function as independent units analogous to gases in a gaseous state.

(The situation begins to become far more complex, however, as the result of a number of circumstances; e.g. expenditure of free solvent with increased concentration, which may result in conditions of shared solvent units and the initiation of forms of "hydrational bonding"; conditions of dissociation and hydration varying with the identity of associated ions; dissociation of radicals incident to osmosis; etc., all of which were found by Dr. Flint to be amenable to mathematical expression through analyses of published specific gravity data and original osmosis experiments.)

On p. 43, Israelachvili 1985, in Table V, lists "(approximate)" hydration numbers and hydrated radius for a number of ions, with qualification that "the total number of water molecules affected is much larger and depends on the method of measurement". Table V is noted to be compiled from data from a number of investigators, including Amis (1975), which was of particular interest to me insofar as I have previously reviewed data from Amis's 1973 book [Amis, E. and J. Hinton, Solvent Effects..., 1973, Tables 3-1, 3-4, 3-15.] Whereas Israelachvili 1985 lists the hydration numbers for Na+ as 4-5 and that for Li+ of 5-6, Amis and Hinton 1973 had listed 14 sets of data from 9 investigators for these same two ions. These values are listed below in Table 1, along with a calculation of the ratios between them, and a calculated average of these ratios.

For convenient comparison, hydration numbers as per Flint are also listed at the bottom of Table 1. Under the circumstance that the hydration numbers listed by Amis are arrived at by various different methods, and may reflect actual differences in numbers depending on the state in which they are measured, the order of agreement between the average of Amis's ratios, .608, and Flint's ratio, .579, seems worthy of note.

TABLE 1:
Hydration numbers per Amis & Hinton 1973; vs. Flint

	Na+	Li+	Ratio
MacInnes	2.0	4.7	.426
"	8.4	14	.600
"	14.9	23	.648
"	9.8	14.3	.685
Washburn	8.4	14	.600
Baborovsky	8-9	13-14	.630
"	9	14	.643
"	44.5	62	.718
Collet	5	7	.714
Haase	13	22	.591
Ulich	2-4	6-7	.462
Devyatykh	10	21	.476

```
        Robinson        5      7    .714
        Gapon           3      5    .600

AVERAGE OF RATIOS                   .608
FLINT (H)              11     19    .579
```

Even more striking is how Gluekauf's relative hydration values [Gluekauf, E., Faraday Soc., Transactions **51** 1241 (1955)] for the series H+, Li+, Na+, K+, Rb+ and Cs+, adjusted to a base value of 11 for Na+ as in Table 2, correlate precisely with Flint's maximum (H+, Li+, Na+ and K+) or zero (Rb+ and Cs+) hydration levels, as shown in Table 2 below. (Please see Table 3, Key, and narrative immediately following Table 3 for further discussion of Flint's determination of "maximum" hydration numbers (H) as exhibited in Tables 1 & 2.)

TABLE 2:
        Gluekauf (X 11/2 =) Adjusted Relative Values

```
H+          3.9                     21 (= H per Flint)
Li+         3.4                     19 (= H per Flint)
Na+         2.0      <<BASE>>       11 (= H per Flint)
K+          0.6                      3 (= H per Flint)
Rb+          0                       0
Cs+          0                       0
```

At a minimum, as shown in Tables 1 and 2, relative hydration numbers derived by various methods and those independently derived by Flint are not incompatible, and might be viewed as providing a measure of mutual validation. In particular, the excellent correlation between the independently-derived relative numbers of Gluekauf and Flint, as in Table 2, would appear to not likely occur by chance.

Noggle 1996, p. 386, notes the historical importance of solution conductivity as "probably the most graphic proof that solutes dissociate into ions ...".

Beyond this, conductivity has also long been regarded as proof that ions in aqueous solution may be hydrated; in 1899, Abegg and Bodländer [Zeit. f. Anorg. Chem. 454-499 (1899), p. 490-1] had noted "One of the first proofs of the existence of such hydrated ions is built on considerations of observed mobilities of ions ...", citing the prior work of Bredig [Zeitschr. phys. Chem. (1894) 13, p. 242, 262]. Bredig, in turn, had referred to the earlier work of Ostwald [Lehrb. d. allgem. Chem. (2. Aufl., Leipzig 1893) **2**, 1, pp. 801, 675-692] as having suggested that ionization is a process of hydration, and as having "indicated that the mobility of elemental ions ... is a distinct periodic function of atomic weight". (At this early date, Bredig had also voiced the suggestion that the respective mobilities of gases in the atmosphere and ions in solution might be mutually explained, citing here Wüllners's Lehrbuch der Experimentalphysik, 4. Aufl. I, 543).

As discussed in my letter of 8 January, Abegg and Bodlander's 1899 suggestion that hydration must vary inversely with atomic weight had sparked Flint's original work in this area. As per Bredig, the idea behind this concept may have originated with Ostwald, whose 1885 publication of the first volume of his <u>Lehrbuch der allgemeinen Chemie</u> has been associated with the very birth of the field of physical chemistry! [E. Bright Wilson, <u>American Scientist</u> **74**, 70 (1986)]

Noggle 1996, pp. 402-416 and 444, provides equations, tables, examples and exercises relating to calculable relationships between transference numbers, mobility, velocity, equivalent conductivity and limiting ionic conductivity for various electrolytes and ions. I assume these were the types of calculations to which you referred in your 29 Jan. letter, e.g.:
--calculation of conductivity from resistance
--total conductivity as the sum of cation and anion conductivity
--equivalent conductance as conductivity divided by concentration
--extrapolation of equivalent conductivity at infinite dilution
--ionic conductivity from equivalent conductivity and   transference numbers
--ionic mobility from limiting ionic conductivity and   Faraday's constant
Without these calculations, specifically those involved in determination of limiting ionic conductivity, Dr. Flint's work would never have existed. As noted above and in my letter of 8 January, the Flint methodology resulted from calculation of hydration numbers from relative conductance; he then utilized the system that was exposed and evolved from this original finding to predict and explain other solute phenomena, with particular emphasis on specific gravity and osmosis, and also including gas solubility, acidity-alkalinity, and other subjects.

In my quest to explore and extend the work of Dr. Flint, I turned his original finding "backwards", so to speak, and sought to examine a wider range of conductance values than discussed by him in his original 1932 report [<u>J. Wash. Acad. of Sci.</u> **22**, 97-119, 211-217 & 233-237 (1932)]. This was the basis for my statement that Dr. Flint's methodology allows for the calculation of conductivities, among other phenomena, which statement in turn apparently obscured the essential distinction between the initial work of Flint and the calculations to which you referred.

Nonetheless, Flint's methodology, itself a quantification of the 1899 observations and suggestions by Abegg and Bodlander concerning the Li+, Na+, K+ and H+ ions and originally derived by Flint from relative mobilities of these ions, enables a set of (relative) conductivity calculations that is not covered in the examples or exercises given by Noggle 1996 (or anywhere in the literature as far as I can tell), which set of calculations does not conflict with and may usefully augment the information presented by Noggle 1996.

As shown in Table 3, observed conductivities for 100% of the 25 mono-atomic positively-charged elemental ions listed in CRC [<u>CRC Handb. of Chem. and Phys.</u>, D167-8 (Chem. Rubber Co., 1985-6)] with Z = 1-19, 37-39 and 55-80 are within 10% of calculated values [relative to Na+ = 50.08], when ions are assumed to be either anhydrous (Zero) or maximally hydrated (Hmax) in accord with Flint's description of hydrational potentiality, and relative conductivities are calculated in accord with Graham's law of diffusion, as first treated by Flint.

The 25 ions listed in Table 3 comprise 68% of the total of 37 mono-atomic positively-charged ions listed in CRC 1985-6; examination of calculations involving the other 12 reveals patterns which may be worthy of further consideration, but which subject gets a bit involved for the discussion at hand. Nonetheless, the agreement exhibited by clear majority (25) of the 37 ions listed by CRC, and 100% of the ions in the groupings Z = 1-19, 37-39 and 55-80 would not very likely occur by chance.

TABLE 3 - Equivalent Ionic Conductivities

Z	C	Obs.(25°C) $(10^{-4} m^2 \ S \ mol^{-1})$		Calc. Error (by Hydrational Assumption) (Hmax)	(Zero)
1 H	+	349.65			-06
3 Li	+	38.66		-03	
4 Be	++	45		08	
11 Na	+	50.08	\<BASE\>	00	
12 Mg	++	53.0		-02	
13 Al	+++	61		03	
19 K	+	73.48		-05	
37 Rb	+	77.8			-09
38 Sr	++	59.4		09	
39 Y	+++	62		04	
55 Cs	+	77.2			09
56 Ba	++	63.6			-08
57 La	+++	69.7			02
58 Ce	+++	69.8			03
59 Pr	+++	69.5			04
60 Nd	+++	69.4			04
62 Sm	+++	68.5			05
63 Eu	+++	67.8			04
64 Gd	+++	67.3			04
66 Dy	+++	65.6			03
67 Ho	+++	66.3			05
68 Er	+++	65.9			05
69 Tm	+++	65.4			05
70 Yb	+++	65.6			06
80 Hg	++	63.6			09

Key:

EQUATIONS 2-5*:

(2) Calculated Conductance = $k/\sqrt{(Z'h)}$; k=527.6

(3) Z' = Z+C, when Z = atomic number and C = valence.

(4) $Z'h = Z' + 9H$, when H = the (maximum) hydration number.

(5) H = 23n - Z', when H = 23 to 0, and n = 1 to 4

      (for Z' = 0 to 23, n=1;

                23 to 46, n=2;

                46 to 69, n=3;

                69 to 92, n=4).

Note*: In his calculations Flint characterized water units involved in hydration as negatively-charged ions (H2O-) with a value of 9, as exhibited above in calculation of Z'h as Z' + 9H.

(Thus 2 x Z'h = 18, the conventionally appraised weight of H_2O.)
   *See Flint, L.H., Behavior Patterns of Hydration, 1964.

   Regarding hydration numbers exhibited in Tables 1 and 2, and atomic numbers and hydration numbers used in calculations exhibited in Table 3, Dr. Flint incorporated a shift in atomic number (Z) equal to one per unit valence (C) into his calculation of what he referred to as the "maximum" hydration number (Equations 3-5 in Table 3, Key); thus, as shown in Table 1, an hydration number of 11 is derived for Na+, as 23 - (Z+C) or 23-12=11. According to Dr. Flint, this shift "permitted an integration of observational data with a satisfying convincing nicety". Without such a shift, the order of agreement exhibited in Table 3, particularly involving the H+ ion, would not have been possible. (More on the H+ ion will follow this discussion.)
   This admittedly controversial action was not taken lightly by Dr. Flint and was discussed in depth in his 1964 book [Behavior Patterns of Hydration, 1964, Institute for Advancement of Science and Culture, New Delhi]. Dr. Flint's discussions included the observations that (1) contemporary science embraces the concept of a shift in atomic-number-equivalent values in the case of an increase in the atomic number of some radioactive elements resulting from the loss of a nuclear electron [e.g., see Bertsch, G. F. and S. McGrayne, in Encyclopedia Britannica, 15th Ed. (1991), Vol. 14, p. 330]; and (2) negatively ionized gases exhibit uniformly higher mobilities than positively-ionized gases, as exhibited in Table 4; thus, in accord with the "Graham principle"/law of kinetic energy, the negatively-ionized gases must weigh less:

TABLE 4:   CONDUCTIVITY OF IONIZED GASES

Ionization:	+	–
H2	5.54	8.45
N2	1.30	1.80
air	1.37	1.81
SO2	.412	.414
C5H12	.385	.451
C2H6O	.363	.373
C2H4O	.307	.331
C2H5Cl	.304	.317
CH3I	.216	.226

(Data from Table 678, SMITHSONIAN TABLES, P. 552)

   H.G. Moseley's definitive correlations of atomic numbers with inverse-square-roots of wavelength may also be cited as a precedent for Flint's use of adjusted values (Z+C) in place of atomic numbers (Z); Moseley's work had involved adjusted values (Z-1) in place of atomic numbers (Z), reportedly "as necessary to fit his data", which adjustment he attributed to the circumstance that "the repulsion of the other electrons cannot be neglected compared with the attraction of the nucleus." [per C.W. Haigh, J Chem. Ed. 72, Nov. 1995].
   Moreover, the fact that Moseley's calculations are regarded as proof of the nature of the atomic numbers, based on direct calculation from empirical measurements (wavelength, or frequency of

vibration) argues for (1) similar treatment of Flint's result in terms of hydration numbers, based on direct calculation from empirical measurements (conductivities); and (2) consideration of Flint's work as a further and independent validation of the primacy of atomic numbers (plus valence), as evidenced in their (inverse-square) relation to conductance - on their own in the case of anhydrous ions, and in conjunction with hydration numbers in the case of hydrated ions.

Following his observation that, for the lighter ions, the hydration number approached zero as the atomic number approached 23, Dr. Flint hypothesized that this pattern, of hydration numbers decreasing from 23 to zero, might repeat beyond Z=23. This led to the characterization of hydrational potentiality as periodic, as per Equation 5 in Table 3, Key. As illustrated in Table 3, calculated values for Sr++ (Z=38) and Y+++ (Z=39) incorporate the assumption of full hydration in accord with Flint's proposed extension of hydrational patterns beyond the lighter elements.

It may be noted in Table 3 that, as per Abegg and Bodlander's 1899 proposal and Flint's 1932 calculation which supported it, the conductance calculation for H+ ion involves the assumption that this ion is anhydrous when placed under the stress incident to the measurement of its electrical conductance. It must be emphasized that (1) this by no means suggests that this is always the case; on the contrary, a fully hydrated state for the H+ ion, precisely in accord with Flint's proposed "description of hydrational potentiality", is evidenced and supported by the relative hydration values of Gluekauf as illustrated above in Table 2; (b) a similar situation may be seen to exist in the case of the SO4-- ion, as discussed in my 1988 abstract entitled "A Lock for Flint: Diameter & Conductivity" [AAAS Pacific Div. Proceedings Vol.7(1988), p. 42], given to you on 5 January, wherein the SO4-- ion may be viewed as fully hydrated in accord with Flint's methodology when measured for diameter, and anhydrous when measured for conductivity; and (c) the general circumstance that ions may exhibit such different hydrational states under differing conditions may be viewed as both agreeing with and possibly helping to explain Israelachvili's statement, (1985) p. 43, that "... the total number of water molecules... depends on the method of measurement."

Overall, it may be offered that the nicety of correlations exhibited in any one of the exemplary sets of calculations discussed above or in abstracts given you on 5 January would not likely occur by chance; taken together, the possibility that all could occur by chance seems quite remote, and at least would appear to comprise sufficient justification for further investigation of the common framework that underlies them.

I am well aware that some of the discussions herein are bound to be controversial, particularly items on the previous few pages. In this already-too-long letter, it has been possible to present only a brief summary of some of the work of Dr. Flint, and myself, that bears directly on these issues; and it has been necessary to gloss over or ignore other very important issues and implications. Nonetheless, I hope this letter has conveyed that the work is sufficiently well-founded and germane to the subjects of conductance and hydration, and possibly other related areas, so as to be worthy of at least a minimum level of consideration by serious students of these subjects. If you agree, I would be interested in any further referral or guidance you might offer which might accelerate this process. If you disagree, I would greatly appreciate learning the reason for such disagreement.

In the meantime, I will continue working toward the placement of an article or articles in appropriate journals as mentioned in my letter of 30 January, with emphasis on the historical perspective as had been suggested by Dr. Prausnitz; in this regard my attempt to communicate with you has served well to apprise me of modern theoretical perspectives in your field and how these appear to relate to my work, and I wish to express my appreciation for that. I may also attempt to establish contact with other U.C.L.A. staff on the possibility of some association through a "History of Science" perspective (as discussed in my 30 January letter) or whose areas of interest may otherwise involve the study of aqueous solution phenomena; if you have any specific recommendations within either of these contexts, based on your reading of this letter, I would appreciate hearing from you.

Thank you again for your kind consideration and attention.

Sincerely yours,
Stuart Hale Shakman

According to my file, we subsequently spoke on February 30, 1996, but notes on this conversation seem to have gone down the "rabbit hole". Nonetheless, the interaction recorded in the above letters seems worthy of passing on, in that they address some of the major "conventional" objections that may be raised relative to Flint's (and my follow-on) work on hydration.

Subsequent to this above letter and our subsequent conversation, the conductance data in the Noggle book provided a simple means of direct calculation of hydration numbers – a near-perfect straight line illustration of the essential "discovery" of Flint in 1932.

This is illustrated above in the item entitled: DIRECT CALCULATION OF HYDRATION NUMBERS FOR LIGHTER IONS, which was submitted to **Nature** 26 Sept. 1996 as proposed scientific correspondence and given a registration number SXA011*, but was subsequently rejected without further explanation.

- - - - -

*Moseley and Harry Jones were mentioned in the above Reiss correspondence without further comment; the below text is included here for historical interest:

MOSELEY

27 Sept. 1995

FLINT AS INDEPENDENT CONFIRMATION OF MOSELEY

Mosely established that the atomic number varies with the square-root of the frequency of vibration, or inversely with the square-root of wavelength. Flint showed how solute ionic number (+ valence) varies with the inverse square of conductance, which seems to be doing somewhat of the same thing from a different direction, i.e., validating the primacy of the atomic number. At the same time, the overall scheme establishes some precise and definite correlation between conductance and frequency, i.e., wavelength varies with conductance to the fourth power, i.e. a four-dimensional correlation wherein the atomic number is sandwiched between inverse squares.. This may be readily illustrated in tabular form:

Table 2: THE FOUR DIMENSIONAL, PONDERABLE, ALGEBRAICALLY CALCULABLE UNIVERSE

	Wavelength	From wavelength to atomic number: Calc At. #				Valence	From wavelength to solute ionic conductivity Calc.Cond.		
	WL	Zcalc= (81/sqrt:WL)+ 7.5	Z-obs	Error		C	EC-calc 545.3/sqrt: ((81/sqrt:WL) +7.5+C)	EC-obs	Error
La	2.676	57.015591	57	-0.000273446		3	70.38878	69.7	-0.00979
Ce	2.567	58.055928	58	-0.000963343		3	69.78653	69.8	0.000193
Pr	2.471	59.028636	59	-0.00048512		3	69.23718	69.5	0.003796
Nd	2.382	59.982455	60	0.000292506		3	68.71091	69.4	0.010029
Sm	2.208	62.011168	62	-0.00018009		3	67.63033	68.5	0.012859
Eu	2.13	63.000285	63	-4.53166E-06		3	67.12164	67.8	0.010106
Gd	2.057	63.976513	64	0.000367122		3	66.63068	67.3	0.010045
Dy	1.914	66.048268	66	-0.000730799		3	65.62346	65.6	-0.00036
Er	1.79	68.042243	68	-0.000620828		3	64.69596	65.9	0.018611

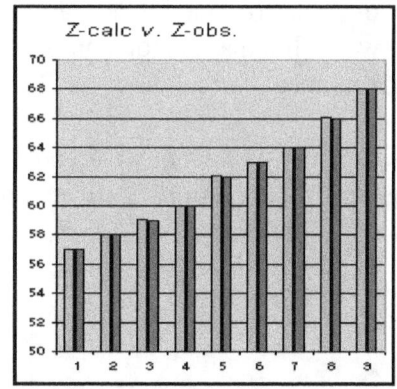

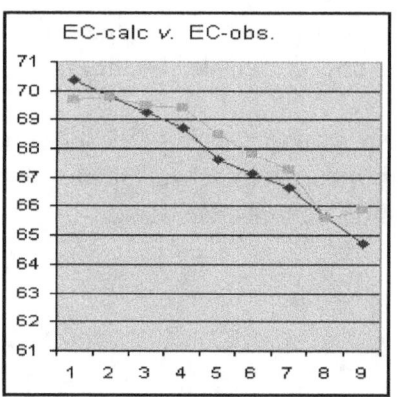

HARRY C. JONES AND HIS HYDRATION OBSESSION

Harry C. Jones, <u>Hydrates in Aqueous Solution</u>, Carnegie Institution of Washington, Publication No. 60, Press of Gibson Bros., Washington, D.C. 1905?

 In his preface, Jones relates Ota had noted, in 1899 while working with "certain double salts" in water, noted that with concentrated solutions "these solutions froze abnormally low; the molecular lowering passing through a well-defined minimum with change in concentration" and that Knight soon after obtained similar results. Jones relates that with three grants from Carnegie, the investigation was extended to about a hundred substances, including acids, bases, salts and neutral organic compounds, with the result showing that this is a <u>general property of solutions</u>. ... [<u>Amer. Chem. J.</u> **22**, 5 (1899) Jones and Ota;
<u>Ibid</u> **22**, 110 (1899), Jones and Knight.
 " ... A possible explanation that occurred to me early in the investigation was that in solution a <u>part of the solvent is combined with the dissolved substance</u>, and no longer plays the rôle of solvent, at least as far as the freezing-point method is concerned. Jones referred to <u>four distinct lines of evidence</u> tht seemed to substantiate this theory, most importantly the unmistakable relation between the lowering of freezing-point and water of crystallization. Also cited were the relation between amount of water of crystallization and temperature, and absorption spectra.
 Jones concluded that an amount of water is combined with the dissolved substance, the amount a function primarily of the nature of the substance, "but for any given compound is a function of the concentration and temperature."

 Jones distinguishes "the present theory of hydrates differs fundamentally from the older theory of Mendeleeff in that according to the latter certain compounds, such as calcium chloride sulphuric acid, and the like, form a few definite compounds with the water in which they are dissolved. According to the present theory, combination between the dissolved substance and water is a general phenomenon; and a given compound, say calcium chloride, forms a complete series of hydrates ranging in composition from a few molecules of water to at least thirty ...."

 As related on p. 2, Jones, <u>Ibid.</u> **23**, 99, 1900, had proposed that in concentrated solutions, chlorides and bromides take up a part of water and that the hydrated chloride or bromide ion acts as a unit or as one molecule in lowering the freezing-point of the remaining water, hut that the total amount of water present is now diminished by the amount that had been taken up.

 And on p. 4, at that stage of the game he noted: "It has been justly said that we do not have a theory of solutions in the broad sense, but only a theory that applies to ideal conditions ... .In a word, we had no theory of concentrated solutions."

FLINT ON THE ATTRIBUTES OF PH, ACIDITY, ALKALINITY

Flint 1933: p. 112 "Jones stated it as an established fact that the velocities of the ions were an inverse function of their mass, while Bredig found that the conductance of an element ion was a periodic function of its atomic weight."

p. 111 "Abegg and Bodlander pointed out that, restricted to salts giving rise to a common negative ion, hydration is inversely related to the atomic weight of the positive ion. This relationship was further corroborated and emphasized in the work of Senter, Bousfield and Jones. Bousfield further pointed out that the specific molecular hydration of the Cl- ion is greater thatn that of the K+ ion, and that the specific molecular hydration of the Na_+ ion is about four times that of the K+ ion."

Bredig, G., Z. Phys. Chem. 13, 191-288 (1894); see relative to periodicity

   For whatever reason(s) Flint abandoned this explanation by 1964. Two intrinsic areas that may have hastened his abandonment of this explanation are
   p. 120 (1) his calculations of the relative acidity of the H+ ion, juxtaposed v. calculations for various negative ions, did not vary as whether the solute involved one, two, or three H+ ion, whereas each charge of the negative ions were significant factoris in calculations.
   p. 122 (2) calculation of conductances of O--, substituted for OH-, provides satisfactory calculated value (172.5 v. 174 observed) but does not account for the effect of the H+ ion which also was derived from the OH- ion.
   Other reasons undoubtedly accounted for Flint's abandonment of the scheme, but nonetheless the nicety of the agreement he had derived is somewhat interesting.

Flint. 1964:Annotations:
Behavior Patterns of Hydration, Ch. 15, "Acid and Alkali Reactivity", p. 130
   To this writer, Chapter 15 was a bit confusing and seemingly redundant.
   For acids, Flint suggested that (10 all cations were hydrated, all anions anyhdrous, and (2) relative acidity was a factor or relative: interdiffusion into air compared with interdiffusion with an associate.
   For alkaloids, Flint suggested (1) all anions and cations are hydrated and (2) relative alkalinity is conditioned (a) qualitatively by relative freedom accorded by the environment and (b) qualitatively by solute ion concentration.

   The imposition of multiple differences in for acidity and alkaline respectively weakens the overall case for the approach as far as it was developed by Flint; however, the extent to which this work did seem to correlate with published data on relative acidity-alkalinity may argue for further investigation along the lines initiated by Flint.

Overall, the suggestion of a relation between mobility and qualities of acidity/alkalinity seem reasonable and some evidence presented by Flint seemed genuinely suggestive.

Ph - and hydration

In 1933 Flint proposed a "tentative acidity-alkalinity index derived from velocity-charge products:, which he compared with observed Ph data, p. 120:

	Flint's Tentative Index	Adjusted	Observed Ph
HCl	5.280	1.720	1.0
H2ClO2	3.87	3.13	3.8
H3BO2	2.375	4.625	5.2
	0.000 = = = = = = = = = = =	7.000]	
NaHCO3	-.923	7.923	8.4
Na2CO3	-2.98	9.72	13.1

p. 118: To determine his index, Flint derived theoret ical velocity values for each ion, multiplied these by their respective charges, divided the larger by the smaller ion, and subtracted one, leaving the derived value negative or positive in accord with which was the larger. A negative value thus was determined to comprise an alkaline substance, a positive value was designated an acid.

ASSIGNMENTS SECTION

Use Flint's methodology to calculate, algebraically from atomic numbers, the following pH values:

Biddle, HC and VW Floutz, <u>Chemistry in Health and Disease</u>, FA Davis Co., Phila, 1969.
--approximate pH values of some common 0.1 solutions of acids and of basic compounds:

acid	pH value	Basic	pH value
Hydrocloric	1.0	Sodium hydrogen carbonate	8.4
Sulfuric	1.2	Sodium tetraborate (borax)	9.2
Acetic	2.9	Ammonium hydroxide	11.1
Carbonic	3.8	Sodium carbonate	11.6
Boric 5.2		Sodium hydroxide	13.2

- - - - - - -   - - - - - - - - - - - - - - - - -

other	
Water	7.0
Gastric fluid	1.5-2.8
Vinegar	3.0
Blood	7.35-7.45

THE PROTON JUMP HYPOTHESIS

That modern investigators have actually been able to devise and exploit an alternate mechanism to understand the H+ ion is a tribute to their ingenuity. However, the goal of a deriving a single active mechanism ...

The protein jump mechanism

Eigen, M. and L. De Maeyer, "Self-dissociation and protonic charge transport in water and ice", p. 505 [Proc. R. Soc. Lond. A 21 October 1958 **vol. 247 no. 1251** 505-533], offers that "the mechanism of structural diffusion provides an explanation of the anomalous H2O+ and OH- mobility and their recombination rate in water."

    p. 511: "the high absolute values of the mobilities of H+ and OH- in water and their anomalous temperature and pressure dependence indicate a special mechanism." and offer "a brief review of the frequent theoretical attempts to explain this mechanism...", from Hückel (1928), who used the proton jump idea dating from Danneel 1905, which "such proton jump mechanism is implied in nearly all the subsequent models", to Bernal and Fowler (1933) to Stearn and Eyring (1937) to Gierer and Wirtz (1949) to Wicke Eigen and Ackermann (1954), Bell (1957), Conway, etal. (1956), concluding that "the H-bond formation (structure diffusion) rather than the proton transfer is rate determining in water", and "based on the assumption that the proton is present in form of the hydronium ion H3O+", <u>with no reference to Flint 1932</u>, whose quantification of the hydrate theory of solutions allows for precise calculation of the mobility of H+ in water on the assumption that it becomes anhydrous under electrical stress.

    p. 509, "H+ and OH- in aqueous solution are strongly hydrated. ... The high value for the hydration energy of H+, which exceeds that of any other univalent ion in aqueous solution by more than 100 kcal/ mole, suggests a strong specific interaction between the proton and a water molecule."
    p. 510, heats of hydration are given as:

H+	Li+	K+	NH4+	OH-	F-	Br-
276	131	92	87	111	94	63

- - - - -

Bernal, J.D., and R. H. Fowler, "A Theory of Water and Ionic Solution, with Particular Reference to Hydrogen and Hydroxyl Ions", <u>The Journal of Chemical Physics</u>, Vol **1**, No. 8, August 1933, 515-548.

p. 516:
  At the outset the authors state "Turning now to ionic solutions, we discuss the nature of ionic hydration. A method of estimating the degree of hydration from the specific gravity of the solutions is developed, and the results compared with those calculated theoretically on the basis of the molecular model. It is concluded that all the strongly polarizing ions H+, Li+, Na+, and all divalent and trivalent positive ions as well as OH- and F- are hydrated, while NH4+, Rb+, Cs+, and most negative ions are not. The degree of hydration depends mainly on the ionic radius ...
  "The H+ ion must exist in solution as (OH3)+. But this makes the anomalous mobility of H+ still more difficult to account for. A break with current theory is proposed. The hydrogen positive ion, effectively

(OH3)+ moves through a potential gradient of 1 volt/cm at a rate 32.5 X 10^{-4} cm/sec. at room terperature. The corresponding rate for (OH)- is 17.8 x 10^{-4} and for all other ions much less (e.g. K+, HH4+, Cl-, 6.7 x 10^{-4}). This discrepancy is analysed and it is claimed that hte excess velocities of these ions over 6.7 x 10^{-4} cm/sec. must be due to a mechanism entirely different from bodily transport through the solution. it is suggested that this different mechanism is the transfer by a jump of one proton from one water molecule to another when favourable configurations are presented. Such an idea has been proposed by Hückel but quite differently developed by him. ..."

p. 517

The authors discuss present theories of ionic mobility, whereby "slower mobility of small ions [is explained] on the hypothesis of a rather indefinite amount of hydration . ...

"However all such theories fail to account for the mobilities of the ions H+ and OH- in water. The equivalent Stokes law radii for H+ and OH- calculated from their mobilities are 2.6 x 10^{-9} cm and 4.8 x 10^{-9} cm, quite impossibly small on physical grounds (and neither hydration nor ionic interaction can do anything but reduce mobility)." The authors note that the study of the anomalous mobility of the H+ ion has led to a "general attack on the nature of water ... ."

p. 531

The authors return to the discussion of the introduction of charged ions into water: "The effect will clearly be roughly proportional to the polarising power of the ion, i.e., its charge divided by the radios. Large monovalent ions will have the least effect, small highly charged ions the greatest. This corresponds to the long familiar hypothesis of the hydration of ions, introduced to account for the apparently anomalous fact that the mobilities of the large ions K+, Rb+, Cs+ Cl-, Br- and I- are all approximately the same whereas small ions such as Li+ or Mg++ move much more slowly. Several experimental methods have been used to find the degree of hydration of ions, but they give widely varying results, and are plainly theoretically unsatisfactory. ...

"The simplest and possibly the most accurate method of estimating the real degree of hydration [532] is the measure of densities of ionic solutions." On the basis of considerations of apparent volume in solution and ionic volume in solids, the authors conclude that the H+ and OH- ions are among those that are hydrated.

p. 534, "Table VI. Extra energy, kcal/ion, derived from coordination of water molecules on ions allowing for the effect of mutual interactions of water molecules. These figures do not represent experimental hydration energies."

Ion	n=1
Li+	15
Na+	11
K+	4.0
Rb+	2.6
R-	4.1
Cs+Cl-Br-I-	<0

In subsequent discussion, the authors state, "The difference between the permanently hydrated Li+, Na+ (K+, F-) ions and the others is not geometrical but physical, the smaller ions carry with them through the water their permanently attached hydration shell; the larger ions on moving exchange water neighbours."

p. 540,

"What has emerged from the study of the molecular structure of water and ionic solution is that it offers no hope of a trivial explanation of the large mobilities of the H+ and OH- ions. If we treat H+ and OH- empirically as the other ions, the densities of their solutions indicate that they are hydrated (see Tables [p.541] IV and V). On the basis of this hydration low mobilities approximating to that of the Li+ ion would be attributed to them. But other properties, particularly viscosity, show that H+ and OH- do not behave normally. The effect of H+ and OH- ions on the structural temperature is very large and negative. This shows that another mechanism than simple hydration must be involved, a mechanism that allows H+ and OH- ions to move rapidly through the solution, and at the same time to impart to it a greater coherence. The second part of this paper is an attempt to describe such a mechanism, and so to explain the observed abnormal mobilities. We believe that the explanation presented is the only one possible, and further possible if, and only if the structure of water is substantially that described in Part I."

Part II. A Quantum Mechanical Theory of the Extra Mobility of H+ and OH- in Water.

"After the discussions of Part * it is impossible to accept the view that hydrogen positive ions are present in aqueous solution as naked protons. They must be firmly attached to at least one water molecule and the whole discussion may proceed on the assumption that the hydrogen positive ion is present in solution in the form OH3+_, the oxonium ion, which like OH- and all other small ions will be more or less hydrated. The extra proton in OH3+ is no doubt accommodated at or near one of the two vacant tetrahedral positions among the electron orbits, the whole structure being very similar to the ammonia molecule NH3.

"Accepting this point of view which very simple considerations of energy are alone enough to demand, the exceptional mobility of this ion becomes all the more striking. If its mechanism of transport through the solution were the same as that of other ions, which necessarily involve bodily transport ... to or towards the proper electrode", it is impossible that its mobility should differ much from that of the ammonium ion ...", concluding that "a special mechanism must be acting both for OH3+ and OH-, the ions of water in water, which is entirely distinct from bodily transport and this view which we shall develop leaves unaffected older views of the transport of any foreign ion through any solution. ...

"... The fundamental idea which must be introduced is quantum mechanical. Having recognised the presence of OH3- structures in the solution, quantum mechanics at once tells us that an OH3+ ion in sufficiently close contact with a water molecule OH2 need not retain its extra proton but can transfer it to the other molecule. The proton jumps to and fro from one to the other when the configuration of the two molecules is favourable ... "

p. 542

"A very similar mechanism must hold for OH-." The authors suggest a proton passes from OH2 to OH-, with the smaller mobility of OH- explained by the deeper position of the H's in OH2 as compared to OH3+.

HYDROGEN BONDS

Krijn, M.P.C.M., and D. Feil, J. Phys. Chem. **1987**, 91, 540-44, "Polarization of the Water Molecule. Analysis of the Changes isn the Electron Density Distribution", discuss an apparent implication that "[OH] bonds are far more polarizable along the O-H directions than perpendicular to them. Remarkable is the significant increase instead of decrease of the electron density in the outer lone pair region of oxygen. ...

... much of the region within 0.75 au from the oxygen nucleus has an electron density distribution that seems polarized in the wrong direction."

- - - - -

Perrin, Charles L., Science **266**, 9 Dec. 1994, "Symmetries of Hydrogen Bonds in Solution", 1665-1668, discuss "simple counterexamples to the hope that the crystal structure reveals the actual molecular structure in aqueous solution". The authors assert that the "disorder of the aqueous environment" may be responsible for a change from symmetric to non-symmetric structure exhibited by monoanions studied. The authors refer to water as "a disorganized medium", and causes hydrogen bonds definitively observed in crystals to be symmetric to alternatively become asymmetric in aqueous solution.

- - - - -

Kendal, James, Smith's College Chemistry, D. Appleton-Century Co., New York 1935, p. 396:
"Ortho-- and Para-hydrogen -- In the hydrogen atom, not only does the electron revolve around the nucleus, but the nucleus has a spin of its own. There are thus two kinds of molecular hydrogen possible, depending on whether the two nuclei are spinning in the same, or in opposite, directions. In accordance with this, it has been found that two kinds of molecular hydrogen do indeed exist, ortho-hydrogen and para-hydrogen. Ordinary hydrogen is a mixture of the two in the ration of three to one. Ortho-hydrogen is slowly converted to para-hydrogen by pressure and cooling, and adsorption on charcoal at liquid air temperature gives practically pure para-hydrogen. The physical properties of the two forms are very slightly different."

- - - - -

Rahman, Aneesur and Frank H. Stillinger, The Journal of Chemical Physics, **55**, No. 7, 1 October 1971, 3336-3357, "Molecular Dynamics Study of Liquid Water" [see also Reiss5 letter], p. 3339:
"Investigation of the structure of ice crystals and the clathrate hydrates is not our primary objective in this paper. But the tetrahedral charge arrangements which underlie $V_{eff}^{(2)}$ strongly favor the local tetrahedral pattern of H bonds about each molecule observed in these aqueous solids."; and p. 3349,
" The internal structure of the water molecule requires that the static structure of the liquid be examined from many more independent points of view, for a given level of comprehension, than is required for liquid argon."; and in the "DISCUSSION" section, p. 3356, "With respect to future study of aqueous solutions ... "One would ... be particularly interested to see if the surrounding water molecules organized themselves into a sort of "cage" analogous to those present to clathrates. The results would be especially interesting in view of the "hydrophobic bond" concept that has been proposed to explain the interaction of nonpolar solutes in water."

BIG BANG, STRINGS AND THE FUTILE SEARCH FOR A THEORY OF EVERYTHING

Rusk, Rogers D., p. 338
Lymon Hanson and Scott, 1951, using 15.7 mV electrons with electron accelerator, "confirmed that nuclear radii are closely proportional to the cube root of the atomic number and that the radius is given to a fair approximantion as $r_o = \sim 1.40 \times 10^{-13} A^{1/3}$cm.
Introduction to Atomic and Nuclear Physics, Second Ed., Appleton-Century-Crofts, N.Y. 1964 (Meridity)

[NOTE: this is 100% in accord with [anhydrous] volume varying as Z, as per Flint]
- - - - -

Maddox, John, Nature
366 297 (25 Nov. 1993), , "Has physics come to an end?", reporting New Delhi University, Maddox offers that "there now seems only a tenuous connection between human-scale laboratory experiments and the grand themes of physics - the structure of the Universe and the nature of the objects, made of particles, that it contains. ...

"Even so, the experiments in the journals are not unconnected with the old grand themes. There is still, for example, much to be done to pin down quantum nonlocality (or to make actual the Einstein-Podolsky-Rosen *gedanken* experiment of 1935) ... . Even the meaning of the concept of a quantum state is now up for grabs experimentally ...

...
"... The most likely outcome of the accelerator experiments being planned is either that these particles [the missing quark and the Higgs boson] will not be found at all or that something else will appear instead. And then, of course, in the most constructive way, the field will come to life again. For a long time, the idea that there is a theory of everything will be forgotten. ... Physics, in other words, is far from dead."
- - - - -

Faye Flam, Science **263** (11 Feb. 1994), 749, reports that CERN's director-general Christopher Llewellyn Smith has indicated that in return for an estimated US$60 million to $100 million a year, CERN's Large Hadron Collider project could accommodate as many as 400 U.S. physicists. This breaks down to about $200,000 per scientist per year. Without additional funding, the U.S. physics community would either have to forego participation or take money from a U.S. project which "could cost people their jobs. 'It comes down to finances and people's livlihood versus science.'" according to Mitch Golding of Harvard. So there you have it - work relief at the cost of $200,000 per. Does this include salary?
- - - - -

"Big Bang not yet dead but in decline", John Maddox, Nature **377** 14 Sept 1995, p. 99, noted that NR Tanvir etal. (Nature **377**, 27-31) had concluded the latest measurements of the hubble constant "implies an age of the Universe much smaller than the known ages of the stars in globular clusters in our galaxy." Madox asserts that this result, "the third of its kind in under a year, makes a nonsense of the standard Big Bang view of how the Universe began."

Maddox relates that "Those with their knife into the Big Bang should also reflect that if the value of H_0 eventually settles down near the values recently published, it will be necessary to come to an understanding of why so many other measurements have given values that are numerically roughly half as big."

Maddox concludes that "The minute it is suggested that these structures [globular clusters] are so exceptional that they must be survivors from an earlier stage in the evolution, the Big Bang will have given way to continuous creation. That will ba a turn-up for the book." It certainly looks like that is what is being suggested!

- - - - -

Oreskes, Naiomi, Kristin Shrader-Frechette, Kenneth Belitz, Science **263** (4 Feb. 1994), p. 641 "Verification, Validation and Confirmation of Numerical Models in the Earth Sciences", begins with the statement "Verification and validation of numerical models of natural systems is impossible" and asserts that "the establishment that a model accurately represents the 'actual processes occurring in a real system' is not even a theoretical possibility."

What does this mean for Flint's methodology and the possibility that it may represent reality? That it is not a model, but rather is a description.

- - - - -

Gary Taubes, Science **269** 15 Sept. 1995, 1511-1513, "A Theory of Everything Takes Shape", notes that in terms of consideration of "string theory", "The sticking point is that the unification of quantum physics and gravity that string theory aims to describe can take place only at an energy scale known as the Planck scale - 18 orders of magnitude higher than the energies achieved by the most powerful accelerators." (that's a million-trillion times higher.)

- - - - -

Frederick Grinnell, Science **272**, 19 April 1996, 333, "Editorial", "Although formal heuristic principles can be helpful in deciding what results might be seen as data, the final outcome will depend on an investigator's experience, intuition, and creative insight."

- - - - -

ADDITIONAL NOTES: HYDRATION/WATER/SOLUTION THEORY, ETC

Meyerowitz, Elliot M., Nature **360**, (3 December 1992), discusses Deng, X. etal in Cell, 791-801 (1992) and cloning of genes involved in a plant's response to light, indicating this work "does not reveal the mechanism of plant phototransduction, but does raise several interesting possibilities", including a possible relation to "evidence that light responses mediated by phytochrome and blue light may act through G proteins ..."
  Article features picture of 4-day old dark-grown mutant v.s. wild-type seedlings, looking kind of like Flint's pix from Smithsonian articles.
- - - - -

Wheatley, DN, Nature **366** (25 November 1993), 308, "Water in life", states "as yet no calculable theory exists with which to begin to consider water's interactions with other molecules."
cites Ling, G.N., A Revolution in the Physiology of the Living Cell (Krieger, Malabar, Florida, 1992)
- - - - -

  Honig, Barry and Anthony Nicholls, "Classical Electrostatics in Biology and Chemistry", Science **268**, 26 May 1995, 1144-1149, note that "much remains to be done before [realizing the goal of] a complete and accurate method to describe the properties of molecules in aqueous solution";
  The authors discuss "the widely used molecular mechanics approach" with its "daunting computational requirements"; and "an alternative approach [which] involves continuum or macroscopic models", particularly "classical electrostatics".
  The authors discuss the use of classical electrostatics as "a successful quantitative tool yielding accurate descriptions of electrical potentials, diffusion limited processes, pH-dependent properties of proteins, ionic strength-dependent phenomena, and the solvation free energies of organic molecules." No calculations are illustrated.
- - - - -

Yam, etal. (1988), p. 330, state that all Stokes-Einstein type equations contain a difficult assumption, "that the mechanisms of mechanical drag are identical with those of global viscosity. Thus for ideal solutions at infinite dilution, the binary diffusion coefficient, D_{AB}, is predicted to be inversely proportional to viscosity. ... [This] suggest[s] that diffusivity is inversely proportional to a variable power of solvent viscosity, proportional to viscosity raised to a power of a function of molar volume, or proportional to a variable power of the solvent molecular weight."
- - - - -

Kanno, H., J. Phys. Chem. 1988, **92**, 4232-6, "Hydration of Metal Ions in Aqueous Electrolyte Solutions: A Raman Study"
p. 4235
  "An inner sphere hydration number change in the rare-earth-ion series is well established.* From the v1 frequency shift due to the hydration number change, we know that hydration enthalpy is greatly dependent on the inner sphere hydration number. Bertha and Choppin** evaluated the hydration enthalpies for all rare-earth members and found that hte hydration ehthalpy rapidly increases from Nd3+ to Tb3+ at which inner sphere hydration number changes. ... [plus other items] These are strong

evidence for the contention that the discrete nature of water molecules will have to be incorporated into a successful solution theory, instead of the continuous dielectric theory for aqueous electrolyte solutions."

*Kanno, H; Hiriashi, J; Chem Phys Lett 1980, **75**, 553; Habenschuss, A, Spedding, FH, J. Chem Phys. 1979, **700**, 2797, 3758; J. Chem. Phys. 1980, **73**, 442.
Bertha, SL, Choppin, GR, Inorg. Chem 1969, **8, 613.
- - -

Born, 1920: p. 45 "Mit Hilfe der von mir kürzlich mitgeteilten Formeln für die Gitterenergie binärer Salze (1) lässt sich, wie Herr Fagans (2) gezeigt hat, die Arbeit berechnen, die man aufwenden muss, um aus der wässerigen Lösung eines Salzes die Ionen in das Vakuum zu beförden."
    [With the help of me recently notified formulas for the lattice energy of binary salts (1) can be, as Mr. Fagan (2) has shown, calculate the work that you have to spend to get out of the aqueous solution of a salt, the ions in the vacuum to beförden]
- - - - -

Kavanau, J. L., Water and Solute-Water Interactions, 1964, Holden-Day, Inc., San Francisco, p. 61, discusses Hindman, J.C., 1962, J. Chem. Phys.**36**:1000: ... his results were incompatible with the concept of a <u>complete</u> hydration shell of at least four tightly bound water molecules (see also Padova, 1963). The hydration numbers decreased with increasing radius of the cation, and there was an approximately linear relationship between the entropy of hydration of the alkali-metal ions and the "effective" hydration numbers (see also Glueckhauf, 1955). For lithium ions he found evidence for the formation of a complete first hydration shell of four water molecules with additional orientation of water in outer shells (even beyond complete orientation of a second hydration shell).
- - - - -
Price, W.E., R. Mills and L.A. Woolf, J. Phys. Chem. **1996**, <u>100</u>, 1406-1410, "Use of Experimental Diffusion Coefficients To Probe Solute-Solute and Solute-Solvent Interactions in Electrolyte Solutions"

Price, etal. 1996, p 1406, discuss "two approaches to the study of interactions in electrolyte solutions using experimental transport coefficients. One is based on generating generalized transport coefficients from experimental data. ... the other approach deals with our attempts to use the water diffusion coefficient and shear visc+osity data to gain insights into ion-solvent interactions and into hydration dynamics of multivalent cations."

p. 1407-8, contrast diffusion coefficients with square-root of ionic strength.
p. 1408, in "Development of a Hydration Model, with reference to secondary hydration relative to "the primary coordinated hydraton sphere", characterize their "less established association" with "more rapid exchange with the bulk water environment". ...
... All the metal ions under consideration [Fe, Cr, Al, La, Zn, Ni, Mg, Zn] have a lifetime for a water molecule coordinated to them that is long compared with the time for a single diffusive step. These water molecules can thererore be considered to move through the solution with the cation as a single kinetic entity."

various numbers discussed in text:

```
Cr3++       12    when it does not form complexes        (19)
Cl-          6    primary hydration                      (7)
Zn(ClO4)2 16                                             (14)
NiCl2       21±2 Ni++                                    (16)
Cr+++       16-17 in CrCl3                               (19)
```

The article concludes with "The use of a value between that of the ion and the bulk water for the diffusion coefficient of the water in the short-lived secondary hydration layer of the cation and the primary anion layer is physically quite realistic in that it allows for sharing of water molecules at high concentrations.

Lee. S.H., and J. C. Rasaiah, <u>J. Phys. Chem.</u> **1996**, <u>100</u>, 1420-1425, "Molecular Dynamics Simulation of Ion Mobility. 2. Alkali Metal and Halide Ions Using the SPC/E Model for Water at 25 degrees C" p. 1420, "The residence times of water in the hydration shells around an ion are found to decrease dramatically with its size." p.1425, in concluding para, refer as statement of fact to "subtle trends in the ion mobility as a function of ion size"

The authors relate that notwithstanding the experimental study of mobilities of ions in water for over a century, theoretical investigations using both continuum and discrete models, and the importance to chemistry and biology, "even the ion mobilities at infinite dilution are as yet incompletely understood."

The authors differentiate between H+ and other ions: "The explanation of the mobility of light ions like H+ in water is a quantum mechanical problem that can be investigated by path integral methods together with ab initio calculations of the intermolecular interactions. [citing here Carr 1985 and Voth 1992] Classical mechanics however is an adequate first approximation for the dynamics of the heavier alkali metal and halide ions in aqueous solution", and discuss problems with classical simulations, including "an accurate but economical representation of the slowly decaying Coulombic interactions in an infinite system and the type of boundary conditions to be used.

p. 1420, indicates that "water molecules in the first solvation shell around the small Li+ ion are stuck to the ion and move with it as an entity for about 190 ps", vs. 35 ps for Na+ and 8-11 ps for large cations.

p. 1423, Table 4, gives residence times of 400 for Li+, 26.4 for Na+ and 9.4 for K+.

[Flint's use of Graham's law recalls similarity between electrical attraction, culoumbic and gravitational.]

Figure 6 contrasts ion mobilities calculated from mean square displacements (MSD) with experimental values. The MSD calculated values and values calculated from autocorrelation functions (VAC) are shown in their Table 5, as listed below and compared with observed values:

	MSD		VAC	observed
F-	5.06±1.63	53.35±17.19	5.02±1.48	55.4
Cl-	6.73±1.13	70.96±11.91	6.65±1.13	76.31

Br-	7.35±0.86	77.49± 9.07	7.27±0.86	78.1
I-	6.54±1.59	68.95±16.76	6.93±1.55	76.8
Li+	4.59±0.35	48.39± 3.69	4.79±1.36	38.66
Na+	4.75±1.83	[50.08±19.29]	5.45±1.75	50.08
K+	7.86±1.75	82.87±18.95	8.45±1.63	73.48
Rb+	8.21±1.79	86.56±18.87	8.56±1.83	77.8
Cs+	7.78±1.36	82.03±14.34	7.74±1.52	77.2

		Cond.	H	1/2	O	Volume	Diam	H	1/2	O
1	H +	373.09			-06		.14			-33
9	OH -	198			06		.25			-26
3	Li +	38.66	-03			34.04	1.29		.08	
11	Na +	[50.08	00]			15.73	[1.00]		.00	
19	K +	73.48	-05			4.93	.68		-.01	
37	Rb +	77.8			-09	4.15	.64			12
55	Cs +	77.2			09	4.255	.65			00
9	F -	55.4		-09		11.55	.90		02	
17	Cl -	76.31		00		4.47	.66	-07		
35	Br -	78.1			-14	4.15	.64			16
53	I -	76.8			05	4.255	.65			02

Note: Ref. 20 refers to M.A. Wilson, A.Pohorille and LR Pratt, J. Chem.Phys. 1985, **83**, 5382. Text refers to this as one of several items (13) which employ molecular dynamics simulations methods.
- - - - -

Hammel, H.T.,J. Phys Chem. **1994**, *98*, 4196-4204, "How Solutes Alter Water in Aqueous Solutions", states "Two theories, one by G. Hulett in 1903 and one by G.N. Lewis in 1908, describe how solutes alter water in an aqueous solution. ... Neither Hulett's nor Lewis's theory incorporates a kinetic theory to account for the altered state of the solvent in a solution. ..."
Hamel notes "the fact that the osmotic pressure of the solvent in a solution is proportional to the absolute temperature".
  p. 4200: states "It is surprising that equations 9 and 11, derived by Lewis, are considered to be fundamental thermodynamic relationships in many standard textbooks of physical chemistry and chemical thermodynamics. Equation 11 is merely a mathematical divice (a gimmic) to calculate changes in chemical potential of solvent in a solution. Equation 99 is not a fundamental relationship even though it and Lewis's concept of solvent activity have become the accepted paradigm for describing the altered state of water in an aqueous solution. ... Unfortunately, Lewis's theory came to dominate the treatment of osmosis and the altered state of the solvent in a solution,
  Hammel concludes that "Disregarding the difficulties of calculation or directly measuring the osmotic pressure of water in an aqueous solution, there is no doubt that an understanding of hte effects of solute on the water were foreseen most clearly by George Hulett. Hulett's theory deserves to be restored to its reightful status in all sciences in which solutions are involved.
  ... [George] "Hulett's theory should supplant Lewis's theory as the preferred account of the altered state of the solvent in a solution."

p. 4201, cites Henry Dixon, professor of botany, Trinity College, Dublin, "the pre-eminent plant physiologist of his time ... and author of the widely accepted cohesion theory." and "his classic monograph entitled <u>Transpiration and the Ascent of Sap in Plants</u>" [Dixon, H.H., MacMillan: London, 1914, p. 140., noting that Dixon is in accord with Hulett fand contrary to Lewis's theory, although he cites neither.

P. 4204, in Appendix A, concludes, "The fact that the osmotic pressure of the solvent in a solution is proportional to the absolute temperature is clue enough to reveal that osmosis is caused by thermal motion and pressure exerted by all molecules at the free surface of the solution and not be thermal motion and diffusion of the solvent molecules through the semipermeable membrane separating solution from pure solvent."

- - - - -

David R. Rosseinsky, Dept. of Chemistry, The University, Exeter EX4 4QD, UK, <u>Nature</u> **360**, 636 (17 Dec. 1992), "Gas correction", qualifies perfect gases as "modeled as comprising point particles having mass but not volume ...", correcting a statement made previously by J. Maddox in <u>Nature</u> **359**, 669; 1992.

D.R. Rosseinsky, same address, <u>Chem. Rev.</u> (1965)dsdss **65**, 467-489, "Electrode Potentials and Hydration Energies. Theories and Correlations", prominently cites the 1907 work of W.R.Bousfield and T.M. Lowry, <u>Trans. Faraday Soc.</u> **3**, 123, in his initial discussion of theoretical procedures, as one of two fundamental factors in calculation of the "Born-charging hydration free energy"

- - - - -

Frank Neville H. Robinson, Senior Research Officer, Clarendon Lab., U. of Oxford, St. Catherine's College, in <u>Encyclopedia Britannica</u>, 1988, Vol. 18, p, 268, "The first thing that is noticeable in a table of ionic mobilities in aqueous solution is that ions differing considerably in mass and size have much the same mobility. Thus, for example, the ions Cl-, Br- and I- have almost identical mobilities although it might be expected that the heavy and bulky I- ion would have a much lower mobility than the lighter and smaller Cl- ion. Even more surprisingly, the small and light lithium (Li+) ion has only about half the mobility of the heavy cesium (Cs+) ion. This is generally attributed to the fact that ions collect a group of water molecules around themselves, which move as a unit with the ion. ..."

- - - - -

Israelachvili, Jacob N. and Patricia M. McGuiggan, <u>Science</u> **241**, 12 Aug., 1988, 795-800, "Forces Between Surfaces in Liquids", note that "The origin of some important fundamental interactions, such as repulsive "hydration" forces and attractive "hydrophobic" forces, are still not understood and offer a challenge for experimental and theoretical work in this area

- - - - -

Wolynes, Peter G., <u>J. Chem. Phys.</u> **68**, 15 Jan. 1978, 473-483, "Molecular theory of solvated ion dynamics", 473, begin:

"The study of transport phenomena in electrolyte solutions has historically been one of the central areas of classical physical chemistry. ....

"It is observed that the mobility of small ions in solutions is benerally smaller than one would expect on the basis of their crystallographic radii and Stokes law. This puzzle is now even more intriguing because of modern evidence for the applicability of hydrodynamics to molecular motion in simple fluids." [citing D.R.Bauer, J.I. Brauman, and R. Pecora, J. Am. Chem. Soc. **96** 6840 (1970), R. Zwanzig and M. Bixon, Phys. Rev. A 2, 2005 (1970)]

Wolynes discusses the "two basically different pictures" that "have evolved in an attempt to understand this problem." The older he refers to as a "solventberg" concept, whereby solvent molecules are rigidly bound to a solvent ion, with the resulting solventberg larger than the original ion; the second is the "dielectric friction" model, first proposed in 1920 by Born and later refined by Boyd and Zwanzig and further developed by Adelman and by Hubbard and Onsager.

Wolynes states that "Neither of these two pictures is totally consistent with experimental observation. ...

"While neither picture is wholly accurate, it seems likely that each contains a kernel of truth. The language of both pictures is quite different, and in the past this has led to considerable argumentation. The proponents of the solventberg picture often point out, correctly, that the dielectric friction picture does not appropriately take into account solvent structure. On the other hand, the proponents of the dielectric friction picture point out the rather too simple dynamical character of the solventberg model."

- - - - -

Israelachvili, J and H Wennerström 1996, portray the "60-year-old issue of hydration forces" as "central to a molecular-level understanding of biological and inorganic systems in water".

- - - - -

ADDITIONAL HISTORICAL NOTES

George E. Kaufmann, California State University, Fresno, Fresno, CA 93740, Journal of Chemical Education **67**, No. 6, June 1990, 451, "Nobel Laureates in Chemistry - A Philatelic Survey, Part I. 1901-1910", lists Jacobus Henricus van't Hoff, Svante August Arrhenius, and Wilhelm Ostwald, the 1901, 1903 and 1909 Nobellists in Chemistry, as founders of physical chemistry. The three joined together in support of Arrhenius's theory of electrolytic dissociation; nonetheless, despite their prominence, "more than a decade of proselytizing ... and a large number of experiments by these three 'ionists' and others were necessary before the theory of electrolytic dissociation was universally accepted."

Kaufmann, JCE **67** July 1990, p. 572, notes Nernst was the 1920 winner, a "German physicist, turned chemist". His heat theorem, eneuciated in 1906, is generally "known as the third law of thermodynamics, which states that the entropy of ordered solids approaches zero as the temperature approaches absolute zero."

Appendix – Gustave LeBon on the Evolution of Matter and Forces

With this Appendix on the great works of Gustave LeBon, we introduce what in "in modern film" parlance what would consider an "prequel". Indeed the great works of Flint do give entrée to the "secret code of the universe", the outline of the precise relation between the essential matter-delimited units of hydration-entities (H2O- ions) and atomic-number-plus-valence-entities (Z+C). And without a doubt, the works of Flint do indeed provide the basis for unifying all that lies scatters in what feebly pretends to so-called "scientific" charades to which we referred to as physics, chemistry and biology, etc.

But undernearth this lies the phememonal edifice exposed by the brilliant researches of Gustav LeBon, that provides the even-more basic glimpses of the evolution of that which comprises the very essence of what we consider "matter" and "energy" and their very orgins. Here we are introduced to the likes of a "ether" with rigidity exceeding that of steel, able to transmit light at a remarkable speed, akin to a "solid without density or weight, however unintelligible this may seem. Here we are introduced to the intermediate world between matter and the ether, the latter consisting of a substance entirely outside the laws of gravitation -- that is to say, imponderable. But at the same time not absolutely incompressible, as indeed from probably such compressibility and condensation, effected at the beginning of the ages by a mechanism totally unknown to us, that are derived the atoms. Thus origin of matter itself could might attributed to condensation nuclei in the ether, having the form of small vortices (or whirlpools) animated with an enormous speed of rotation. And within this edifice, LeBon reserves a special place for the magic of hydration. In his words …

Gustav LeBon
Evolution of Matter – Evolution of Forces
Notes/ excerpts with reference to hydration and origin of phosphorescence and radioactivity
Hydration in EoM 154, 179, 386, 390; EoF 364, 374

EVOLUTION OF MATTER – 1902

"'The secret of all who make discoveries' says Liebig, 'is that they look upon nothing as impossible'" [Gustav LeBon]

xxii F. Legge 1906 (Translator's Preface) "In France, even more than in England, it has always been considered an impertinence for any one not engaged in the tuition of youth to possess original ideas on any scientific subject, and the violence of Dr. Le Bon's adversaries was only equaled by the volubility with which they contradicted themselves and each other."

p. 6 "The aptitude of matter to disaggregate by emitting effluves of particles analogous to those of the cathode rays, having a speed of the same order as light, and capable of passing through material substances, is universal. The action of light on any substance, a lighted lamp, chemical reactions of very different kinds, an electric discharge, etc., cause these effluves to appear."

p. 12 ... various forms of energy resulting from the dissociation of material elements, such as heat, electricity, lo the ether."

p. 19, Chapter II, History of the Discovery of the Dissociation of Matter and of Intra-Atomic Energy: "My researches preceded in their beginning, all those carried out on the same lines. It was, in fact, in 1896 that I caused to be published in the *Comptes rendus de l'Academie des Sciences*, solely for the purpose of establishing priority, a short notice summing up the researches I had been making for two years, whence it resulted that light falling on bodies produced radiations capable of passing through material substances. Unable to identify these radiations with anything known, I pointed out in the same note that they must probably constitute some unknown force -- an assertion to which I have often returned. To give it a name I called this radiation black light (luminiere noire).

"At the commencement of my experiments I perforce confused dissimilar things which I had to separate one after the other. In the action of light falling on the surface of a body there can be observed, in fact, two very distinct orders of phenomena:"
1. Radiations of the same family as the cathode rays. They are incapable of refraction or of polarization, and have no kinship with light. These are the radiations which the so-called radio-active substances, such as uranium, constantly emit abundantly and ordinary substances freely.
2. Infra-red radiations of great wave-length which, contrary to all that has hitherto been taught, pass through black paper, ebonite, wood, stone, and in fact most non-conducting substances."

p. 22-3 In 1897 LeBon via experiments "arrived at the conclusion that the radiations of uranium were not in any way polarized", thus "not any form of light, but an absolutely new thing ... a new force: 'The properties of uranium were therefore only a particular case of a very general law.' It is with this last conclusion that I terminated one of my notes in the comptes rendus de l'Academie des Sciences of 1897.

"For nearly three years I was absolutely alone in maintaining that the radiations of uranium could not be polarized. It was only after the experiments of the American physicist, Rutherford, that this came to be accepted. ...

p. 25   "When the question as to polarization was definitely settled, it took but little time to establish the correctness of the facts as stated by me. But it was only after the German physicists, Giesel, Meyer, and Schweidler, discovered, in 1899, that the emissions of radioactive bodies were, like the cathode rays, capable of deviation by a magnet, that the idea of a probable analogy between all these phenomena began to spread."

p. 27 "I have had the satisfaction of seeing, while still alive, the recognition of the facts on which I based the theories which follow. For a long time I had given up all such hope, and more than once had thought of abandoning my researches. ... Several of the notes sent by me to the Academy of Sciences provoked absolute storms. ...

p. 28 "Today, when my ideas have slowly filtered into the minds of physicists, it would be ungracious to complain of their criticisms or the silence of most of them towards me. Sufficient for me is it that they have been able to avail themselves of my researches. The book of nature is a romance of such passionate interest that the pleasure of spelling out a few pages repays one for the trouble this short decipherment often demands. I should certainly not have devoted over eight years to these very costly experiments had I not at once grasped their immense philosophical interest and the profound perturbation they would finally cause to the fundamental theories of science. ...

p. 39 "... astronomers [elucidate] that the condensation of our nebula suffices by itself to explain the constitution of our solar system. It is conceivable that an analogous condensation of the ether may have begotten the energies contained in the atom. The latter may be roughly compared to a sphere in which a non-liquefiable gas was compressed to the degree of thousands of atmospheres at the beginning of the world."

! ! ! analogy to compression of gases at floor of ocean to form life !!!

p. 40   "The different methods in use for measuring the speed of the particles of dissociated matter, whether radium or any metal whatever, have always given nearly the same figures. This speed is almost that of light for certain radio-active emissions. For others we get a third of that speed.

p. 51   The methods of dissociation are, as we shall see, numerous. The most simple is the action of light. It has further the advantage of costing nothing. In so fresh a field, with a new world opening out before us, none of our old theories should stop those who seek. "'**The secret of all who make discoveries' says Liebig, 'is that they look upon nothing as impossible'** ... The power to dissociate matter freely would place at our disposal an infinite source of energy ... . The poor would then be on a level with the rich, and there would be an end to all social questions."

LEBON DISCUSSES IMPLICATION OF DISSOCIATION/EXPLOSION
p. 77 "The scholar who discovers the way to dissociate instantaneously one gram of any metal -- radium, lead, or silver -- will not witness the results of his experiment. The explosion produced would be so formidable that his laboratory and all the neighouring houses, with their inhabitants, would be instantaneously pulverized. So complete a dissociation will probably never be attained, though M. de Heen attributes to explosions of this kind the sudden disappearance of certain stars. Yet there is hope that the partial dissociation of atoms may be rendered less slow. I assert this, not as the result of theory, but as of experiment, since, by means set forth in the sequel, I have been able to render metals almost deprived of radioactivity, like tin, forty times more radio-active than an equal surface of uranium."

p. 78 [from *Annee Scientifique*, 47th year, pp. 6, 88, and 89: "M. Gustave Le Bon was the first, as we should not forget, to throw some light into this dark chaos, by showing that radioactivity is not peculiar to a few rare substances, such as uranium, radium, etc., but is a general property of matter, possessed in varying degrees by all bodies. ..."]

[M. Sagaret, in *Revue Philosophique*, November 1905: "No scientific theory has responded nor can better respond to our yearning for unity than that of Dr. Gustave Le Bon. It sets up a unity than which it would be impossible to imagine anything more complete, and it focuses our knowledge on the following principle: one substance alone exists which moves and produces all things by its movements. This is not a new conception, it is true, for the philosopher, but it has remained hitherto a purely metaphysical speculation. Today, thanks to Dr. Gustave Le Bon, it finds a starting point in experiment.

" ... The atom, like the living being, is born, develops and dies, and Dr. Gustave Le Bon shows us that the chemical species evolves like the organic species."

LENARD: DISSOCIATION OF SODIUM; RELATE TO INTERDIFFUSION CALCS?

p. 85 [The] transport (entrainement) of matter is ... observed in most electrical phenomena, notably when electricity of a sufficiently high potential passes between electrodes. The spectroscope, in fact, always reveals, in the light of the sparks, the characteristic lines of the metals of which these electrodes are composed. Yet another reason seemed to prove the material nature of these emissions. They could be deviated by a magnetic field, and were therefore charged with electricity. Now, as no one had yet seen the transport of electricity without material support, the existence of such a support was considered evident.

"The sort of material dust which was thus supposed to constitute the emissions from the cathode and those from radio-active bodies presented singular characteristics for a material substance. Not only does it present the same properties whatever the body dissociated, but it has also lost all the characteristics of the matter which gives it birth. Lenard showed this clearly when he sought to verify one of his old hypotheses, according to which the effluves generated by ultra-violet light striking on the surface of metals are composed of the dust torn from those metals. Taking sodium, a body very easily dissociated by light and the smallest traces of which in the air can be recognized by the spectroscope, he found that the effluves thus emitted contained no trace of sodium. If, then, the emissions of dissociated substances are matter, it is matter which has none of the properties of the substances whence it comes.

"Facts of this nature have multiplied sufficiently to prove that in the cathode ray radiation, as well as in radioactivity, matter transforms itself into something which can no longer be ordinary matter, since none of its properties are preserved   "So long as the existence of this intermediate world was ignored, science found itself confronted with acts that it could not classify. ...

p. 89 Lord Kelvin ... "considers the ether to be 'an elastic solid filling all space.'

LeBon on the ether:
Evolution of Matter, p. 89

"Unfortunately the properties of the ether do not permit it to be in any way likened to a gas. Gases are very compressible and the ether cannot be so. If it were, in fact, it could not transmit, almost instantaneously, the vibrations of light. It is only in theoretically perfect fluids, or, better still, in solids, that distant analogies with the ether can be discovered, but then a substance with very singular qualities has to be imagined. It must possess a rigidity exceeding that of steel, or it could not transmit luminous vibrations at a velocity of 300,000 kilometres per second. One of the most eminent of living physicists, Lord Kelvin, considers the ether to be 'an elastic solid filling all space.' But the elastic solid forming the ether must have very strange properties for a solid, which we never met with in any other. Its extreme rigidity must be accompanied by an extraordinarily low density -- that is to say, one small enough to prevent its retarding by its friction of the movement of the stars through space"

p.90 ... one is thrown back on the idea that **the ether is a solid without density or weight**, however unintelligible this may seem."

NOTE: IF THE ETHER IS ANALOGOUS TO WATER, WHY IS EARTH SLOWING? OBVIOUSLY THERE IS SOME SORT OF RESISTANCE, ARGUABLY UNITS OF ETHER?

p. 91 Le Bon states the ether "must has mass, since it offers resistance to movement. The mass is slight, since the speed of the propagation of light is very great. If there were no mass, the propagation of light would probably be instantaneous."

p. 92 Lord Kelvin is cited as having concluded "that the ether consists of a substance entirely outside the laws of gravitation -- that is to say, imponderable. But he adds, 'We have no reason to consider it as absolutely incompressible, and we may admit that a sufficient pressure would condense it.'"

"It is probably from this condensation, effected at the beginning of the ages by a mechanism totally unknown to us, that are derived the atoms, considered by several physicists -- Larmor especially -- as condensation nuclei in the ether, having the form of small vortices (or whirlpools) animated with an enormous speed of rotation."

POINT OUT ANALOGY TO ORIGIN OF LIFE!

p. 95 " ... The osmotic equilibria which control most of the phenomena of life are revealed by the attractions and repulsions of the molecules in the bosom of liquids. The movements of various substances and the varieties of equilibrium resulting therefrom thus play a fundamental role in the production of phenomena. They constitute their

essence, and form the only realities accessible to us."... ...
"vortices [appear]

p. 96 vortices appear, theoretically at least, to play a
preponderant part. Larmor [Ether and Matter, 1900] and
other physicists consider that electrons, the supposed
elements of the electric fluid -- and according to some
scholars, of material atoms -- are vortices or gyrostats
formed within the ether. Professor de Heen [Prodromes
ad'un Theorie de l'Electricitie] compares them to a rigid
wire twisted into a helix, the direction of their rotations
determining the attractions and repulsions. Sutherland
seeks in the direction of the movements of these gyrostats
the explanation of the electrical and thermal phenomena of
conduction. 'Electric conduction,' he says, 'is due to the
vibration of the gyrostats in the direction of the electric
force, and thermal conduction to the vibration of vortices in
all directions.' ["The Electric Origin of Rigidity"
Philosophical Magazine, May, 1904]"

p. 96 discusses the fundamental role of vortices in the ether
as substantiated by their role in reproduction of in
attractions and repulsions in electrical phenomena, in the
constitution of material substances with geometric forms.
"A material vortex may be formed by any fluid, liquid or
gascous, turning round an axis, and by the fact of its
rotation it describes spirals."

P. 97 "These vortices constitute one of the forms most
easily assumed by material particles, since a fluid can be
caused to whirl by a simple breath. The can produce,
besides, all the movements of rotation, and very stable
equilibria capable of striving against the power of gravity
as a top in motion remains upright on its pivot. It is the
same with a bicycle, which falls laterally when it ceases to
roll forward. The helices with vertical axes called
helicopters used in certain processes of aviation rise in the
atmosphere by screwing themselves into it so soon as they
are put in rotation ... . It will thus be easily conceived
that it is in rotatory motion that is found the best
explanation of the equilibria of atoms."

p. 104 "The cathode rays ... simply represented, in the
original theory of Crookers, molecules of rarefied gas,
electrified by contact with the cathode, and launched into
the empty space within the tube at a speed they could never
attain of they were obstructed, as in gases at ordinary
pressure by the impact of other molecules ... [but] p. 105
"Measurement of the electric charge of the particles and of
their mass has proved that they are a thousand times
smaller than the atom of hydrogen, the smallest atom
known. One might doubtless suppose in strictness, as was
done at first, that the atom was simply subdivided into

other atoms preserving the properties of the matter whence
they came; but this hypothesis broke down in face of the
fact that the most dissimilar gases contained in Crookes'
tubes gave identical products of dissociation, in which were
found none of the properties of the substances from which
they had issued. It had then to be admitted that the atom
was not divided, but was dissociated ... ."

p. 107 Le Bon asserts that the ionization of saline solutions
and ionization of a simple body, e.g. a gas, are wholly
different things, insofar as the former yields variously
different components depending on the combination
present in the compound, whereas the former always yields
an identical product, i.e., ions or electrons identical in all
bodies.

p. 108 thus in Le Bon's world, "the term ionization applied
to a simple body merely means dissociation of its atoms,
and not anything else."

p. 116-117 Le Bon takes issue with the prevailing view: i.e.
that neutral atom loses negative electrons that are
surrounded by neutral molecules and become the negative
ion and (3) the original atom, with an excess positive
charge, attracts neutral particles and forms the positive ion.
Says Le Bon, p. 117, "Things, however, only happen in the
manner described in a gas at ordinary pressure. In a
vacuum, electrons do not surround themselves with a
retinue of material molecules; they remain in the state of
electrons and can acquire a great speed; so that the
formation of negative ions is not observed in a vacuum.
Nor does the positive ion in a vacuum surround itself with
neutral particles, but, as it is composed of all that is left of
the atom, it is still voluminous, which is why its speed is
comparatively feeble."
    Lebon continues to state that as "in radioactive bodies,
that the negative electron are expelled from the atom into
the atmosphere, at the ordinary pressure, with too great a
speed for their attraction on the neutral molecules to be
capable of exercise. ... It is they that form the Beta rays of
Rutherford."

p. 120, regarding electrons:
    "Their apparent mass -- that is to say, their inertia -- is,
as we shall see in another chapter, a function of their speed.
It becomes very great, and even infinite, when this speed
approaches that of light.

p 124, regarding the speed and size of cathode particles:
"... As the force necessary to deviate to a given extent a
projectile of known mass enables us to determine its speed,
it will be conceived that it is possible to deduce from the
extent of their deviation the velocity of the cathodic

particles. It is seldom less than one-tenth of that of light, or say 30,000 kilometres per second, and sometimes rises to nine-tenths.

p. 154, regarding phosphorescence of sulphate of quinine: "After seeking the cause of its phosphorescence on cooling, and proving it to be due to a very slight hydration, I noted that by reason of this hydration the substance became radio-active for a few minutes. It was the first instance I discovered of the dissociation of matter -- that is to say, of radio-activity -- by chemical reactions, and it led me to the discovery of many more."

p. 161 "We have just seen that very different causes acting in a continuous manner, such as light, can dissociate matter and finally transform it into elements which no longer possess any material properties, and cannot again become matter. "This dissociation ... must have played a great part in natural phenomena. It is probably the origin of atmospheric electricity, and no doubt that of the clouds, and consequently of the rainfall which exercises so great an influence on climate. One of the characteristic properties of radio-active emissions is that of condensing the vapour of water, a property which also belongs to all kinds of dust ..."

p. 178 Le Bon notes that whereas it was commonly thought that the instability of the likes of uranium and radium "was a consequence of the magnitude of their atomic weight", his researches show "that it is just those metals whose atomic weight is feeblest, such as magnesium and aluminium, which become most easily radio-active under the influence of light' in contrast to heavier elements like gold, platinum and lead. "Radio-activity is therefore independent of atomic weight ... "

p. 179 "... Salts of quinine ... are not radio-active. By letting them be slightly hydrated after desiccation, they become so, and remain phosphorescent while hydration lasts. ..."

...

"The disintegration of matter necessarily implies a change of equilibrium in the disposition of the elements which compose the atom."

p. 186 Regarding the phosphorescence of salts of radium, it "is lost by the action of heat and only reappears after the lapse of a few days. Humidity destroys it altogether."

P. 202 "... Given that a considerable liberation of energy results from the dissociation of a very slight quantity of matter, **the creation, in the future**, of such a machine -- that is to say, **of an apparatus giving forth a power**

extremely superior to that expended in setting it in motion -- **can be considered possible**."

p. 231 "The scientific revolution now going on seems rapid, but this rapidity is much more apparent than real. The transformation of present ideas on the constitution of matter, which seems to have taken only a few years, was prepared, in reality, by a century of researches.

"Scientific ideas, in fact, only change with extreme slowness, and when they seem to be abruptly modified, it is always noted that this transformation is the consequence of a subterranean evolution which has taken long years to accomplish.

"Five fundamental discoveries form the bases on which have been slowly built up the new ideas relating to the constitution of matter. They are -- 1st, the facts revealed by the study of electrolytic dissociation; 2nd, the discovery of the cathode rays; 3rd, that of the X rays; 4th, that of the bodies called radio-active, such as uranium and radium; 5th, the demonstration that radioactivity does not belong exclusively to certain bodies and constitutes a general property of matter."

p. 236 "I have spoken in a former chapter of the millions of corpuscles per second which one gram of a radio-active body can emit for centuries. Such figures always provoke a certain amount of mistrust ... [which] disappears when one notes that very ordinary substances are capable without undergoing any dissociation, of being for tears the seat of an emission of abundant particles easily verified by the sense of smell, without this emission being discoverable by the most sensitive balances."

p. 236-7 Lebon discusses attempts by M. Berthelot "to determine the loss of weight undergone by very odoriferous though slightly volatile bodies. ... and he has arrived at the conclusion that one gram of idoform only loses the hundredth of a milligram of its weight in a year: and that the comparable weight loss for musk would be very much smaller. ... [and] 'that "there is hardly any metallic or other body which does not manifest, especially on friction, odors of its own ... '"

p. 242 shades of Van't Hoff: "The equilibria determined by the attractions and repulsions which are born in the bosom of solid bodies, are discernible with difficulty, but we can render them visible by isolating their particles. The method is easy, since it only consists in dissolving the solids in some suitable liquid. The molecules are then nearly as free as if the body were transformed into gas, and it is easy to observe the effects of their mutual attractions and repulsions. **It is well known, moreover, that the molecules of a dissolved body move within the solvent**

and develop there the same pressure as if they were converted into gas in the same space."

...

p. 242-3   "It is in the phenomena called osmotic that molecular attractions and repulsions are most clearly shown." Le Bon continues to discuss how water poured into a solution of a colored salt solution such as sulphate of copper, the molecules diffuse themselves into the liquid. "There consequently exists in them a force which enables them to overcome the force of gravity. This force of diffusion is the consequence of the reciprocal attraction of the particles of water and of the dissolved salt."
ANTI-GRAVITY IN SOLUTIONS!!
... "Osmotic attractions are very energetic. In the cells of plants they can make equilibrium to pressures [p. 244] of 160 atmospheres, and even more according to some authors."

p. 246   "'Osmotic pressure' says Van't Hoff, 'is a fundamental factor in the various vital functions of animals and vegetables. According to Vries, it is this which regulates the growth of plants; and, according to Massart, it governs the life of pathogenic germs.'"

p. 258-9   "The mineral [259] being is characterized by its crystalline form as the living being is characterized by its anatomical one. The crystal also undergoes, like the animal or the plant, a progressive evolution before attaining its final form. Again, like the animal or the plant, the mutilated crystal can repair its mutilation. The crystal is in reality the final stage of a particular form of life."

p. 259   "... Among the secretions of every microbe there always appear, according to him [Schroen] , crystals characteristic of every species of microbe.

"These observations show that during its pre-crystalline period -- that is to say, its infancy, the future crystal behaves like a living being. It represents tissue in course of evolution. It is an organized being undergoing a series of transformation of which the final stage is the crystalline form, as the oak is the final stage of the evolution of the acorn. The crystal would therefore seem to be the last phase of certain equilibria of matter unable to rise to the forms of higher life."

p. 260   "Far from being an exceptional state, the crystalline form is, in reality, the one to which all forms tend, and which they attain so soon as certain conditions of the medium are realized. Salts dissolved in an evaporating solution, and a melted metal when cooled, always tend to assume the crystalline form; and if we consider, as we do nowadays, that solutions show close analogies with gases,

it may be said that the two most usual forms of nature are the gaseous and the crystalline."

p. 261   "Observation shows that all living beings from bacteria up to man, always proceed from an earlier thing."

p. 261-2   "It may ... [262] be said that crystals present two very distinct modes of reproduction -- spontaneous generation and generation by affiliation.

"This faculty of spontaneous generation, possible to the crystal being, is, as is well known, impossible to the living being. The latter is only produced by affiliation, and never spontaneously. However, it must be admitted that before being born by affiliation, the original cells of the geological periods must have been born by affiliation, the original cells of the geological periods must have been born without parents. We are ignorant of the conditions which permitted matter to organize itself spontaneously for the first time, but nothing indicates that we shall always be thus ignorant."

p. 274-5   "'It is certain that in the future as in the past,' writes M. Lucien Poincare, **'the most profound discoveries, those which will suddenly reveal regions entirely unknown, and open up perfectly fresh horizons, will be made by a few men of genius who will pursue in solitary meditation their stubborn labour, and who, to verify their boldest conceptions, will doubtless require only the most simple and least costly methods of experiment.'**

"Considerations such as these should always be borne in mind by independent seekers when the find themselves stopped from want of means, or by the indifference or hostility which most often requites their labours. ... It was thus that the attentive study of the effluves generated by light on the bit of metal struck by it was the origin of all the researches noted in this work ... .

"The great interest of such researches, when stubbornly followed up, consists in constantly seeing new facts appear, and in never knowing into what unknown region one will be led. I have noticed this more than once during the many years devoted to my experiments."

p. 276   "To obtain the transformation of certain bodies we shall require no energetic means, such as high temperatures, great electric potential, or the like."

p. 280   "It is known that, according to the theory even then old but greatly developed a few years ago by Arrhenius, in an aqueous solution of a salt -- chloride of potassium, for example -- the atoms of the chloride and of the potassium separate and remained present in the bosom of the liquid. Chloride of potassium is dissociated by the sole fact of its solution into chlorine and potassium. But, as potassium is a metal which cannot remain in water without violently decomposing it, nor find itself in presence of chlorine without energetically combining with it, **it must perforce be admitted that the chlorine and the potassium of this solution have acquired new properties bearing no analogy to their ordinary properties.** It follows from this that their atoms have been entirely transformed. ... [This] would lead us to consider the atom as the easiest thing in the world to transform, since it would suffice to dissolve a body in water in order to obtain a radical transformation of its characteristic properties."

p. 284   "[The] principle of the transformation of the properties of a substance by the addition of a very small quantity of another body has thus plainly a general importance."

p. 285   "The various applications I have made of this principle have proved to me that it will be fruitful and of practical use, not only in chemistry and physiology, but also in therapeutics. I base this assertion on some studies which I undertook several years ago on the totally new properties caffeine assumes when associated with very small doses of theobromine (an alkaloid which, when isolated, only acts on the organism in very large doses). From experiments made on various patients, several of which have been repeated in one of the laboratories of the Sorbonne by Professor Charles Henry, theobromized caffeine would seem to be the most energetic muscular stimulant known. Observations made on a certain number of artists and writers have likewise proved its singular power on intellectual activity."

p. 299 .. Colloidal platinum or gold are certainly no longer either gold or platinum, thought made from these metals.
    "The properties of colloid metals have, in fact, no analogy with those of a salt of the same metal in solution. ...
    "The properties, at once so special and so energetic, of these metals led perforce to the study of their action on the organism, which is very intense. It is to their presence in various mineral waters that Professor Garrigou attributes several properties of these waters -- that of abolishing the phenomena of intoxication, for example. M. Robin has employed colloid metals as a remedy for sundry affections, notably typhoid fever and pneumonia, by injecting from 5 to 10 cubic centimetres of a solution containing 10 milligrams of metal per litre. The result was a considerable increase of the organic exchanges, and of the oxidation of the elimination

[300] products as revealed by an over-production of urea and uric acid. These solutions being, unfortunately, very rapidly alterable, their practical use is very difficult."

p. 301   "At the time when bacteria were believed to constitute the active agent of intoxications, it was possible to explain by their rapid multiplication the intensity observed in their effects, but it is now known that the toxins remain just as active after the bacteria have been separated by filtration. The living substance called yeast transforms glucose into alcohol and carbonic acid, abut after having killed this yeast by heating it to a certain temperature, a substance can be extracted from it deprived of all organisms and called zymase, as capable of producing fermentation as the living yeast itself. The phenomena attributed a few years ago to microorganisms are therefore due to non-living chemical substances fabricated by them."

p. 302   "Among the substances of which one might strictly say that they act only by their presence is found the vapour of water, which, in extremely small doses, plays an important part in various reactions." LeBon procedes to discuss how for example "dry acetylene is without action on hydride of potassium, but in presence of a trace of humidity the two bodies react ... with such violence that the mixture becomes incandescent." and other examples.

p. 307 "The doctrine of its [the atom's] immutability reigned for two thousand years, and nothing allowed us to suppose that it might one day be shaken."

p. 319 "It is in these atomic universes, whose nature was so long misunderstood, that must be sought the explanation of most of the mysteries which surround us. The atom, which is not eternal as the ancient creeds asserted, is far more powerful than if it were indestructible and therefore incapable of change. It is no longer a thing inert, the blind sport of all the forces of the universe. These forces, on the contrary, are its own creation. It is the very soul of things. It stores up the energies which are the mainspring of the world and the beings which animate it. Notwithstanding its infinite minuteness, the atom perhaps contains all the secrets of the infinite greatness."

- - - - - - -- - -
Experimental

p. 321 "The scientific or philosophical doctrine which has not experience for its basis is deprived of interest and constitutes only a literary dissertation without meaning".

p. 386 hydrations: The dissociation of matter is observed in many reactions, and especially in hydrations. Oxidations, even the most energetic (oxidation of sodium in moist air, for instance), have generally little or no action.

p. 390 "The phenomena of radio-activity -- that is, the emission of effluves -- obtained with uranium, thorium and radium, are very noticeably modified by heat and moisture. ... As to hydration, it suppresses phosphorescence and diminishes radio-activity.

"The diminution of the action on the electroscope by hydration varies greatly with the substances employed." This is followed by data/ examples of discharge for uranium nitrate, uranium oxide, thorium oxide and radium bromide.

"I should add that if the water acts chemically, it at the same time acts partly by the absorption of a part of the emitted particles -- that is to say, like a screen.

"Wetted, or simply exposed to moisture, radio-active bodies lose all phosphorescence, which is not at all the case with ordinary phosphorescent bodies, and they only regain it when brought to a white heat."

p. 394 |"It was in gases that the dissociation of simple bodies was first observed, and that at a time when one hardly thought of speaking about the dissociation of atoms. The phenomenon was then called by the name of ionization. This term, in reality, should be considered as absolutely synonymous with that of dissociation of matter, as I have already stated."

p. 414, on aluminium -- "M. Ditte has concluded, from his numerous experiments, that aluminium is a metal easily liable to attack under many conditions, several of which are still undetermined. The fact appears indisputable. The Navy has been compelled to abandon the use of aluminium, and unless means be found to associate it with a metal able to modify its properties, it will be impossible to employ it, as has been proposed, for metallic constructions."

p. 426, Lebon notes that before his 1897 published experiments, "It was absolutely unknown that metals struck by light acquired properties identical with those of the uranium and the cathode rays."

REFERENCES TO GIANTS ON HYDRATION
Arrhenius 280
Bredig 297
Davy 184, 231
Van't Hoff 246
Lenard 31, 85, 112, 127, 345, 371, 372, 422, 426, 427
Ostwald 261, 281, 282
Van der Waals 256

Gustave Le Bon
The Evolution of Forces (London 1908)

12 Le Bon, regarding solar heat, recites "that dissociation appeared sufficient to explain the maintenance of the sun's temperature"; and regarding electricity, he recalls results of his experiments which showed "that the particles emitted by an electrified point are identical with those which come forth from a radio-active body such as radium. This fact proves that electricity also is a product of the dematerialization of matter."

24 "Forces are known to us solely by the movements they generate. Mechanics, which claims to be the [25] foundation of the other sciences and to explain the universe, is devoted to the study of these movements."

"On [the] notion of the invariability of inertia, or in other words, of the mass, are based the edifices of chemistry and mechanics."

30 Le Bon discusses variability of mass: mass varies by dissociation of atoms, and further, that products of dissociation have mass varying with velocity.
"... not only does the mass vary by the dissociation of atoms, but, further, that the products of this dissociation have a mass varying with their velocity."

35 Speaking of the contemporary physics establishment, and discrepancies between experiments and extant formulas that sought to explain the phenomena, "So soon as the equation no longer agreed with the experiments, they [the involved mathematicians] rectified the equations by imagining the intervention of 'hidden movements', which completely baffled observation. ..."

"Notwithstanding such subterfuges, the insufficiency of the classical mechanics has every day become more manifest with the progress of physics." ...

"It is not wholly in the great questions relating to the synthesis of the universe that the classical mechanics has shown itself very insufficient, but also in apparently much more modest problems like the theory of gases. It is by invoking the calculation of probabilities, by imagining a kind of statistics, that it arrives at establishing extraordinarily complicated and also extraordinarily uncertain equations, which elude all verification."

40 "The critical mind is so rare a gift that the most profound ideas and the most convincing experiments exercise no influence so long as they are not adopted by scholars enjoying the prestige of official authority.

"Nevertheless, it always happens in the long run that a new idea finds a champion in some scholar possessing this prestige, and it then rapidly makes its way. ... "

42 Le Bon discusses the three principles of thermodynamics, noting regarding the first, that "The quantity of energy contained in the universe is invariable.", concedes that if defined as the sum of visible and potential being constant, this "remains unassailable, because ... [43] we can always attribute to [potential energy] the value necessary to satisfy the required ratio."

42 Regarding the second principle of thermodynamics, "the principle of Carnot" as enunciated by Clausius, "Heat cannot pass from a cold body to a hot without work.", is generalized as "The transport of energy can only be effected by a fall in tension. ... It is applicable ... to all forms of energy -- calorific, thermal, electrical, or otherwise.
"This passage of energy from the point where its tension is highest to that where it is lowest is perfectly comparable to the flowing of a liquid contained in a vessel communicating by a tube with another vessel placed at a lower level. ..."

44 Of Carnot: "His genius-inspired idea was just to compare a fall of heat to a fall of water, and all subsequent progress has consisted in recognizing that the various forms of energy, electricity in particular, obey, in their distribution, the laws which regulate the flow of liquids. ..."

51 "The product of the quantity by the tension, that is to say, the work, is a constant magnitude; but it is possible, without altering that product, to increase one of the factors and to diminish the other. ..."

55 Regarding matter, Le Bon asserts we "have simply shown that it constitutes a form of energy capable of transforming itself into other forms, and that it is, through its dissociation, the origin of most of the forces of the universe, notably solar heat and electricity. ..."

56 "... Observation indeed shows that the disappearance of any form of energy is always followed by the apparition of a different energy; but this evolution is accompanied by a degradation of the original energy, which becomes less utilizable. The sole exception to this is perhaps gravific energy."

60 mass increases with velocity as proved by experiments with radio-active particles; indicates that the ether must offer resistance as proved by the non-instantaneous nature of the propagation of light.

60 "Many hypotheses in physics, such as the kinetic theory of gases, would probably quickly vanish if experiment could throw light on them."

61 LeBon says proof of loss of energy is dim, but points the way

64-5 In discussion asserting the difference between energy and work, i.e. criticizing physics for defining all dissimilar forms of energy "as equivalent to a certain amount of mechanical work", Lebon states "We should have [65] a very poor idea of the comparative value of a horse, a negro, and a white man, if we confined ourselves to measuring the number of kilommetres that each could produce. Little can be know of things from simply measuring one of their quantitative elements. ...", quoting Ostwald as having arrived at the same conclusion.

83 Le Bon discusses the circumstance of endless cycles of evolution in the universe, and untold numbers of civilizations come and gone. Speaking of the known hundreds of millions of stars and nebulae and unknown numbers beyond, "Their past must be of fearful length … During these accumulations of ages unknown to history … the millions of stars … must have begun or ended cycles of evolution analogous to that now pursued by our globe. Worlds peopled like ours, covered with flourishing cities filled with the marvels of science and the arts, must have emerged from eternal night and returned thereto without leaving a trace behind them."

Of the very large "numbers of luminous stars, planets, and nebulae existing in the firmament, "Spectrum analysis shows that they are at very different stages of evolution. Their past must be of fearful length, since geologists estimate the existence of our planet at several hundred million years.
"During these accumulations of ages unknown to history, the millions of stars with which space is peopled must have begun or ended cycles of evolution analogous to that now pursued by our globe. Worlds peopled like ours, covered with flourishing cities filled with the marvels of science and the arts, must have emerged from eternal night and returned thereto without leaving a trace behind them."

107 "Adopting the theoretical ideas put forward by Clausius, Arrhenius recognized that an electric current was in no way necessary to produce the dissociation of compounds into ions. In dilute solutions, the bodies dissolved must be separated into ions by the mere fact of

solution. When the [108] electrodes of a battery are plunged into such a solution, the ions must simply be attracted by them -- the positive ions by the negative pole, and the negative ions by the positive pole."

...

108 "This chlorine and this sodium in the state of ions must differ much from the substances generally known by those names, since the sodium of our laboratories cannot be introduced into water without decomposing it. ...

...

108 "All these theories and the experiments whence they are derived show us that electricity is every day more and more considered as the essential factor in the properties of bodies."

108   Valence
109   Affinity

108-9 Referring to the current theory that "the aptitude of ions to unite with a greater or less number of other atoms of various [109] bodies -- must depend on their capacity for saturation with electricity.", Lebon states   "Such is the current theory. It is very probable that things happen in a less simple, perhaps even in a very different manner; but when an explanation fits in fairly well with known facts, it is wise to be satisfied with it."

126. Le Bon asserts the division between static and dynamic electricity "weighed down science for over 50 years."

127 Le Bon asserts that "the electricity generated by a battery is identical with that produced by a static machine." Why do the terms differ? "The textbooks are silent."

129-30 "colossal energies are contained in the smallest particle of matter... play perhaps a preponderant part in biological phenomena. ... we evidently find ourselves here in presence of a new world."

142 Le Bon remarks "in passing that Hertzian waves are to electricity what radiant heat is to that circulating in matter. ... The causes which determine the production of waves of radiant heat seem to possess only distant analogies with those of electric waves."
... . Hertzian wave: electricity :: radiant heat: heat circulating in matter

"The more we study matter, the more we are struck by its extraordinary sensitiveness. Under its apparent rigidity it possesses a very complicated structure and an intense life."

144 "The discovery by M. Branly of the variation in conductivity of metal filings under the influence of the Hertzian rays, although it passed unnoticed when first published, certainly constitutes one of the most remarkable discoveries of modern physics ... . This is that the fragments of metal in contact with one another and presenting considerable resistance ... [145] ... to the passage of electricity, become conductors under the influence of very weak electric waves, and lose this conductivity by a simple shock."

146 mentions use of Wheatstone bridge to measure resistance
146-7 Ohm's law and copper

147-8 "We have seen that Hertzian waves propagate themselves in space to a distance of hundreds of kilometres, and produce on the metallic bodies they meet electrical induction currents capable of manifesting themselves in the form of sparks. [148] " ... unfortunately, by reason of the size of these waves, and the consequent phenomena of diffraction, the greater part of the energy is lost. To concentrate it, mirrors of gigantic dimensions would be necessary. Discusses giant mirrors to capture herziaan waves ... [Oudin Resonator] ...   "However, I do not believe the solution of the problem is impossible ..."

149 The End of War Machine -- "The problem of sending a pencil of parallel Hertzian waves to a distance possesses more than a theoretical interest. It is allowable to say that its solution would change the course of our civilization by rendering war impossible. The first physicist who realizes this discovery will be able to avail himself of the presence of an enemy's ironclads gathered together in a harbour to blow them up in a few minutes, from a distance of several kilometres, simply by directing on them a sheaf of electric radiations. On reaching the metal wires with which these vessels are nowadays honeycombed, this will excite an atmosphere of sparks which will at once explode the shells and torpedoes stored in their holds.
   "With the same reflector, giving a pencil of parallel radiations, it would not be much more difficult to cause the explosion of the stores of powder and shells contained in a fortress, or in the artillery parks of an army corps, and finally the metal cartridges of the soldiers. Science, which at first rendered wars so deadly, would then at length have rendered them impossible ... ."

153. walls and hills are transparent to herzian waves.

157 electric waves pass readily through very narrow slits, while square openings large enough for the insertion of a finger will not allow them to pass." They act like metal

disk which passes through a slit, "but is stopped by an orifice smaller than its diameter."

158 Electric waves, being very large, easily pass round large obstacles; while light rays, being very small, can only pass round obstacles of the dimension of a hair. ... It is solely because he could not observe it, that Newton contested the wave theory of undulations. ..."

161 water absorbs electric rays

167 particles of radioactive bodies are projected into space at nearly the speed of light
"The rotary movements attributed to the elements of matter alone explain how the particles of radio-active bodies are projected into space with a velocity of the same order of light.

"These velocities of rotation are necessary not only to explain the projections just mention, but also the equilibrium of the elements of which the atoms are formed. In the same way as the top falls to the ground the moment it ceases to turn, the elements of matter only keep themselves in equilibrium by their movements. If these last were stopped for a single instant, all bodies would be reduced to an invisible dust of ether, and would no longer be anything."
...

"In my last book we considered electricity as an intermediary substance between matter and the ether, [168] resulting from certain disturbances of equilibrium of the ether following upon the partial disaggregation of the atoms. ...

"If, therefore, matter can, by dissociating itself, produce various energies -- light, heat, electricity, etc. -- this does not in any way say that it is composed of light, heat, or electricity. The conception of electrons, a near relative to the old phlogiston, is, as has been well shown by Professor de Heen, one of the most unfortunate metaphysical ideas recently formulated." ... [i.e., the concept of the electron is flawed; origin of force is disturbance of equilibrium]
"The notion of a disturbance of equilibrium as the origin of any kind of force, is fundamental ... in order to understand the various forms of energy. Particles at rest are no more electricity than the ether at rest is light. When the equilibrium of this ether is disturbed, it experiences certain vibrations which we call light. ... It is the same with electricity. As soon as the particles [of matter (SHS comment)] constituting fluid termed electric are no longer in equilibrium, then, and only then, appear the phenomena called electricity. ..."

"All forms of energy being produced by disturbances of equilibrium, it results from this that by varying these disturbances we generate different forces."
KEY TO INFINITE POWER E.G. AS PER MORAY / TESLA !!!

Of "the various forms of electricity", Le Bon posits [169] 1. The Electric Fluid, [170] 2. Static Electricity, 3. Dynamic electricity, 4. Magnetism, 5. The electric atoms or electrons, 6. Cathode rays, 7. X-rays, 8. Negative Ions, [171] 9. Positive Ions, 10. The Ionic Fluid, 11. Neutral Electricity (presumed non-existent), 12. Electricity condensed in chemical compounds, 13. Electric waves.

170-1 ions
175 "We are able to measure heat without knowing anything of its essence."

176 Le Bon observes that attempts to explain heat by laws of mechanics/thermodynamics is a failure, referring to early physicists' phlogistic theory, followed by an imponderable, otherwise similar, fluid to be absorbed or expelled, which nonetheless enabled Sadi Carnot to "have thought of comparing the flow of heat to that of a liquid... "

178 "'The analogy existing between the propagation of heat in athermanous bodies and the filtration of fluids through porous masses is so close," writes M. Boussinesq ..., 'that we might seek to obtain from it a mechanical theory of conductivity if there were such a thing as a caloric fluid.'
"We are not certain, moreover, that this fluid does not exist. ... For the moment our ignorance on this point is complete."

178 "The effects of heat on matter are of daily observation. ... a [slightest] variation in temperature ... suffices to modify its electric resistance ... The slightest oscillation in the ether causes it to vibrate and radiate. There is thus a continuous exchange of energy between matter and the ether."

183 "Heat being very early known, ... it was natural to take it as the unit of measurement. ... But it now appears that some very active energies can be manifested under other conditions than heat, and cannot, in consequence, be measured by it. The temperature becomes less and less as we advance towards the extreme ultra-violet [where] ... it is only perceptible to instruments of excessive sensitiveness."

184 absolute zero is "deduced from the consideration that, as gases contract by 1/273 of their volume per degree, at 273 degrees below the ordinary zero they could contract no

further." ... but "If heat be only the consequence of the movements of the particles of matter, as is generally admitted, [185] these movements would cease at the absolute zero. ... [as would], no doubt the other forces, such as cohesion."

186 Le Bon explains that "a heated body does not radiate anything resembling heat." Rather, "it produces vibrations of the ether [which] ... being affected by the molecules of the air or the bodies placed before it, generate heat. These vibrations are not heat, but simply a cause of heat ... " light is radiant heat!

188, note 1 "... If the importance of a scholar is measured by the consequences of his works [principle of conservation of energy], it might be said that Mayer was one of the 5 or 6 greatest men of his century. ... The official professors, who saw the principle of Mayer daily growing in importance, could not accept the fact that so considerable a discovery had not issued from their own laboratories, and united their efforts to try and efface from the annals of science the great name of Mayer. [e.g.] ... Dr. Tait, Professor of Physics at Edinburgh wrote 'Mayer, by a lucky chance, came across a method which has turned out a good one.' ... This epithet of 'lucky chance' is, however, freely applied to those who discover anything. [e.g.] ... a **Cambridge physicist [who] .. recognized that the universal dissociation of matter which I had made known was 'the most important theory of modern physics,' but, he added, I had only discovered it by a 'lucky guess.' All the merit was due to the specialists who had taken steps to check its accuracy."**

189 "A vessel of polished metal loses its heat by slow degrees ... . The same metal, if covered with lacquer ... rapidly parts with its heat."

190 "Until the absolute zero is reached, matter unceasingly sends vibrations into the ether."

190 night vision

194   "Light is produced by vibrations of matter propagated under the form of waves in the ether. When these waves possess a length suitable for impressing the eye, we give them the name of visible light. ...
... We ought to give the name of light to the visible or invisible radiations emitted by matter ... [as with] ... radiant heat.
   "Matter, then, is incessantly transformed into light at all temperatures. ..."

202 " ... the position of the maximum of calorific energy is found in the luminous part of the spectrum. But as, in comparison with the total length of [the infra-red], the visible region is of very small extent, it follows that the total calorific energy is much greater in the invisible infra-red. According to the last measurements of Langley, the visible solar spectrum only contains one fifth part of the calorific energy of the infra-red region. ...
   "This immense invisible region ... ought to play a very important, though hardly suspected, part in the phenomena of vegetable life and in meteorology."

203 Compared to the 1/5 of solar radiation that is visible, Le Bon contrasts that no observers of artificial light have found a loss of less than 90 percent.

204 Le Bon asserts that "the discovery of a means of transforming the invisible calorific energy into visible light would effect" would yield great savings for artificial lighting.

208 "There is no body entirely transparent to all radiations. ...
   "We now attempt ... to explain the transparency and opacity of bodies by a phenomenon resembling that of acoustic resonance. ... Matter may be considered to be composed of small molecular tuning-forks capable ... of vibrating to certain notes, but not to others. ... [209] The radiation that causes them to vibrate remains, in issuing from the transparent body, exactly the same as when it entered ... .
   "This theory ... is only maintainable if we supposed the molecules of bodies to be already animated by rapid movements ... .

210 ... in the case of opacity, the plate is heated and then emits in all directions ... radiations with a wave length greatly differing ...   "If the amplitude of the luminous waves striking an opaque body be great enough, the molecules of this body may be driven sufficiently apart to cause it to pass into the liquid or gaseous state. [liquid and gaseous states – beyond – ethereal? Ionic?]

218-9 Lenard - wrong interpretation - discharge of electricity by metal struck by light is not from pulverization but rather by dissociation of its atoms.

219 "Nothing would be more instructive for the history of evolution of ideas than the recital of the uncertainties through which those engaged in research have passed, and of which their final works naturally contain no trace." For this reason Le Bon points "out the causes of error which delayed me for a long time ... "

223 " ... All bodies at a high temperature, such as are apparently those which exist in the interior of our planet, liberate torrents of electric particles analogous to those produced by radium. I do not know whether they serve to maintain the earth's heat, but it seems more reasonable to believe that they play a part in the production of earthquakes."

223 " ... the energy emitted by the dissociated body in the form of different particles may be far superior to the energy which provoked its dissociation. Light, then, acts like a spark on a mass of gunpowder"

225 phosphorescence-knowing animals may be more numerous than sunlight-knowing
   Le Bon discusses four "classes" of phosphorescence: 1. that generated by light, 2. independent of light and determined by different physical excitants e.g. heat, friction c., 3. by chemical reaction, including life (eg. fireflies), and 4. invisible to the eye.

231 results enabling resolution of the question of existence of antagonistic action

233 how to tell Cape diamonds from Brazilian: "Cape diamonds ... are always inferior to the Brazilian, not only by their hardness but also by their brightness. ..."
Following illumination by a bright ribbon of magnesium, "Nearly all the Brazilian and all of those of the Bahia mine were .. brightly phosphorescent ... [but] Not one of the Cape diamonds was phosphorescent."

241 an invisible opposing force
   "The general appearance of [a curve representing the loss of phosphorescence over time] ... shows ... that things happen as if the reaction exciting the phosphorescence placed itself in equilibrium with an opposing force ... . When this equilibrium is attained, the reactions ... stop completely."

246-7 discussed many compounds phosphoresced by heat, "most of these are of aqueous origin"

255 Le Bon discusses experiments that "prove definitely" that phosphorescence by heat and that by light are phenomena of the same order.

263-4 hydration

264   "The only clearly defined reaction which allowed me to obtain phosphorescence is hydration." referring to experimental details discussed in Evolution of Matter.

266-7 discusses phosphorescence of living beings, with the bottom of the sea "covered with veritable forests of phosphorescent polyps ..."
267   Bacteria and Phosphorescence

270 99% of incandescence is useless for light; rather it is dark heat; the mercury (phosphorescent) lamp gives 40% for light; the output of phosphorescent animals is higher e.g. the glow-worm higher still, with "nearly the whole of energy ... transformed into visible light. Phosphorescence will probably be the artificial light of the future"

271 "Phosphorescence as a manifestation of intra-atomic energy" ... represents the transformation into light of various energies of which only a few are known."

274 discusses intra-atomic origins of phosphorescence, with "the complete solution of the problem ... still far off. ... the reactions producing phosphorescence ... are of a special order, for which the laws of the early chemistry are of no service. How ... could for example, the very slight hydration of a salt of quinine ... act on so stable a structure as an atom, and at the same time generate both radio-activity and phosphorescence?"
274 Hydration
274 Intra-atomic origin of phosphorescence

277 introduces the idea that chemical reactions can be produced within solid bodies, e.g. diamonds, as manifest in phosphorescence.

278 "We are hardly in reality beginning to suspect the causes of phosphorescence; but that which we catch a glimpse of enables us to feel by anticipation that it will constitute one of the most important chapters of chemistry ... and ... the history, as yet hardly outlined, of the dissociation of matter."

279 black light = invisible light
   "The appearance in 1896 of the work of Roentgen on the x-rays determined me to publish immediately ... a note on some particular radiations capable of passing through bodies, which I had been studying for 2 years ... . I called them by the name of Black Light by reason of their sometimes acting like light while remaining invisible."
   Le Bon later found they "were composed of 3 very distinct elements: (1) radio-active particles of the family of cathode rays; (2) radiations due to invisible phosphorescence."

280 He designated by the name of Black Light the latter 2 types. Regarding those of invisible phosphorescence

281 "they seemed so astonishing to physicists that they preferred not to believe them. Yet the verification of the most fundamental of them was extremely easy, and demanded no other expense than 50 centimes in money and a few minutes in time. I now know, however, that several repeated them, but, astonished at their success, preferred not to speak of results which ... official science had not consecrated ... ." Le Bon refers to M. Gariel of the Faculte de Medecine de Paris as having acknowledged "'These facts are almost extraordinary'" yet dismissed them because "'phenomena relating to radiations are certainly not yet all known.' ...

"The invisible phosphorescence which I discovered is characterized by the following phenomena: (1) A phosphorescent body exposed to the light preserves for a period of about 18 months the property of emitting invisible radiations capable of refraction and polarization, and of impressing photographic plates. The spectrum of these radiations, which is analogous to that of light, only differs from it by its invisibility. (2) At the end of these 18 months, the body has no longer any appreciable radiation, but preserves indefinitely a residuum which can be made visible by projecting on its surface dark infra-red radiations."

283 "There exist two forms of invisible phosphorescence" that following the visible and that preceding it. "They can both be easily transformed into visible light."

...

"The majority of bodies struck by light preserve sometimes for a very long time the property of emitting dark radiations capable of impressing photographic plates. But it is with those capable of first acquiring phosphorescence that the phenomenon can be best studied.

"First, here are the experiments by which I determined the properties of the light thus emitted."

284 "1. Duration of the emission and variation in the intensity of the rays emitted as a function of the time." wherein calcium sulphide powder between 2 strips of glass is exposed to sunlight for a few seconds and then placed in absolute darkness. After three days, when placed on a photo negative over a gelatino-bromide plate, "a very vigorous image of the negative is obtained in two hours." after 15 days, 12 hours; 25 days, 30 hours; 6 months, 40 days; 18 months, 60 days.

"2. Propagation in a straight line, and refraction. –" as demonstrated with "A statue coated with sulphide of calcium dissolved in copal varnish ... exposed for a few seconds to the light. Three or four days after it has become entirely dark, ... [285] using a portrait camera with large aperature, we obtain, by exposures varying from eight to 15 days, images as perfect as those taken in daylight."

"Polarization. –" demonstrated using "A strip of Iceland spar ... introduced into the optic system of the camera previously used" "with two glass tubes in the form of a cross, filled with sulphide and fixed ... in a good focus ... beforehand.

286 "This experiment proves, at the same time, the emission of invisible radiations, their propagation in a straight line, their refraction, and their aptitude for polarization.

"4. Composition of the rays emitted. – The perfect sharpness of the images obtained ... proves that the index of refraction of the lens for dark rays is the same as for visible light."

287 Le Bon concludes that the "dark light" differs from solar light only by its invisibility, due to the small amplitude of its waves, and that the residual invisible light lasts for a long time.

291 Discusses an experiment where dark radiations added to others yield light; almost converse of Fresnel's light+light=dark

316-7 Discusses concept of using reflections of invisible light to view enemy targets in the dark

318   "Down to the absolute zero of temperature, all bodies incessantly radiate, as has been seen, waves of light invisible for our eyes, but probably perceptible by the animals called nocturnal ... .

"... There does not exist in nature, in reality, any dark bodies, but only imperfect eyes. All bodies whatever are a constant source of visible or invisible radiations, which, whether of one kind or the other, are always radiations of light."

318 Human sensing device

319 "As the invisible infra-red rays form the greater part of the solar spectrum, it may be imagined that they play a considerable part in meteorology and in vegetable physiology."

320   Bacteria and Phosphorescence

326 Discusses experiments with lettuce seeds and light; anticipates Flint, discussing trials: "Lettuce, cucumbers, grains, etc., set to germinate under blass bellglasses, covered with black paper, transparent to the infrared, all germinate quicker than by the light of the sun, and then wither and die in about a fortnight. ...

327 refers to light baths in medicine

344-5 gravity - speed of propagation must be immensely higher than the speed of light (per Laplace), perhaps (per Poincare) on the order of light vibration's velocity

345 "We do not know how gravitation is propagated, but it seems to me that the law of the inverse square of the distance allows us to imagine gravific waves having a form analogous to that of the waves of light, electric waves, etc. It is, in fact, only to forces which are propagated in this way that such a law is applicable. …" spherical waves of gravity

347 "All of the planets of course act upon one another, which is why Kepler's laws are approximately correct.

349 notes that if the earth's rotation were 17 times as fast, the centrifugal force would annul gravitation at the Equator.

351 antigravity and levitation; "Gravitation is only an attraction, which can be annihilated by a corresponding repulsion, as that exercised on masses of iron can be annulled by the action of a magnet."

352 materialization and other psychic phenomena: Lebon clearly distances himself from the likes of the so-called N-rays:
"As to the so-called psychic forces, materializations, etc., it will be useless to busy ourselves with them here. They have attracted the attention of eminent scholars, such as Crookes, Lodge, Richet, and others, but they have yet to be demonstrated, and until this is done, it is better to try to interpret the phenomena observed by known causes. I had occasion to examine without prejudice, and with the assistance of M. Dastre, a subject with a European reputation, but our investigations, continued throughout several séances, disclosed to us nothing demonstrative. The story of the N-rays, moreover, shows us the difficulty of thorough observations in similar matters, and of avoiding causes of error. Before building a temple to unknown forces, we ought to be perfectly certain that they do not issue wholly from that domain of illusion in which all divinities have hitherto been born."

353 LeBon discusses the response by Chemists as to causes of cohesion, affinity, catalysis, osmosis, crystallization, diastasis -- "All these phenomena belong to the cycle of molecular and intra-atomic energies, complete acquaintance with which is reserved for a science much more advanced than our own.
"The most constant and most easily observed effects of these forces are the attractions and repulsions which take place between the different elements of matter."
"We have seen that matter is composed of infinitely small particles gravitating round one another as the planets round the sun, and probably formed by whirls in the ether. Matter is ether already organized, having acquired certain properties such as weight, form and permanence."

354 the origin of all phenomena is in attractions and repulsions, reciprocal as with a sprin g

356 solutes act as gases, analogous: "the smallest quantity of gas, introduced into a receiver, spreads into all parts of it. … To compress a gas … a considerable pressure must be exercised upon it.
"In solution, the molecules of the dissolved bodies behave in a similar way. They exercise upon the walls of the vessel containing them a pressure called osmotic, which is easily measured. The solution of a body has for this reason been compared to a gas."

357 discusses force of electro-magnet to counter gravity as means of viewing ether in a form able to exercise strong attractions

358 notes that "Two different saline solutions with the same osmotic pressure, have the same freezing point; while the vapour-pressures of liquids are equal at temperatures equally removed from their boiling point, and bodies of the same atomic weight possess the same calorific capacity, etc." Cites Van der Waals "law" that all fluids have the same properties in corresponding states.

359 crystallization – liquids solidify into geometric shapes – why? Governed by unknown forces. Asks: "Why … do liquids when solidifying take certain regular geometrical forms …? The hidden causes of the form of a crystal are as unknown to us as those of a plant or an animal. These things happen as if physico-chemical phenomena were governed by unknown forces which compel them to act in a predetermined direction.
"We, however, may in some sort comprehend the genesis of a few forms by reducing them to extremely simple cases – for instance, to molecular attractions within a liquid.
"When we introduce into an aqueous solution a drop of liquid of different osmotic pressure, the molecules of the two liquids are attracted or repelled, and sometimes for fairly regular figures. We may also, by combining

attractions and repulsions of electrical origin, obtain greatly varied figures. …"

360 artificial plants: LeBon relates the demonstration of a combination of chemicals yields an artificial "seed" which germinates into an artificial plant, illustrating that osmotic equilibria may condition certain external forms, likens demonstration to difference between man and statue.

362 life of cells

363 activities of cells, chemically:
LeBon discuss how "That which the [the forces of life] are accomplishing every instant of our existence soars far above all that the most advanced science can realize. The scholar capable of solving by his intelligence the problems solved every moment by the cells of the lowest creature would be so much higher than other men that he might be considered by them as a god."

364 "Life is only maintained by an incessant using up of the materials borrowed from outside. …

365 role of intra-atomic energy in life
  -- dissociation of matter within the organism; radioactivity in breath, per Prof Dufour
  -- excitants (cola and coffee) increase in energy due to dissociation of atoms (physiologists have not yet explained)

367 discusses forces which regulate the organism and others that direct their force: regulating and morphogenic forces --   autonomic (regulating forces)

368 seemingly relates directed vital force as following from concepts of pleasure and pain

369 quotes M. Gustave Bonnier on the tremendous complexity of a fragment of protoplasm, and thus "not more difficult to create afresh a living animal – an elephant for instance – than to create a speck of living matter."

371 quotes M. Dastre: "The naturalists give us nothing but words … names which are applied to collections of facts.
  "These terms … are … not entirely vain words if we use them to simply translate facts of observation instead of considering them as explanations."

373 mentions the new theory of abrupt mutations and implications on Darwinian

375 LeBon asks "Is it possible to pass from the laws of crude matter to those of living matter? .. Up till now we have found no bridge capable of linking together these two orders of phenomena"